Anatomy and Physiology
A Self-Instructional Course

4. The Respiratory System &
The Cardiovascular System

Ralph Rickards
M.A., D.Phil. |
Marketing Director of |
Cambridge Consultants (Training) Limited

David F. Chapman
M.B., F.R.C.S.(Eng.), F.R.C.S.(Ed.)
Senior Registrar, Otolaryngology
Royal Berkshire Hospital, Reading

CHURCHILL LIVINGSTONE
EDINBURGH LONDON AND NEW YORK 1977

CHURCHILL LIVINGSTONE
Medical Division of Longman Group Limited

Distributed in the United States of America by Longman Inc.,
19 West 44th Street, New York, N.Y.10036 and by associated
companies, branches and representatives throughout the world.

© Longman Group Limited 1977

ISBN 0 443 01669 0

Printed in Singapore by
Singapore Offset Printing (Pte) Ltd.

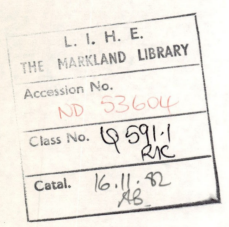

CONTENTS

THE RESPIRATORY SYSTEM

THE CARDIOVASCULAR SYSTEM Pages 19-75

1. THE ANATOMY AND PHYSIOLOGY OF THE RESPIRATORY SYSTEM

1.1. Introduction

All cells take up oxygen (O_2) which reacts with simple compounds within the mitochondria of the cell to produce energy-rich compounds, water, and carbon dioxide (CO_2). The energy-rich compounds are consumed in the energy-using activities of the cell. The exchange of oxygen and carbon dioxide between the cells of the body and the environment is called *respiration*.

Oxygen is carried to the tissues, and carbon dioxide is carried from the tissues, in the blood.

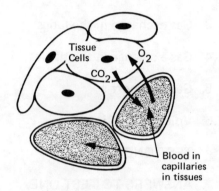

The function of the respiratory system is to allow the uptake of oxygen from the air into the blood, and permit carbon dioxide to be lost from the blood into the air.

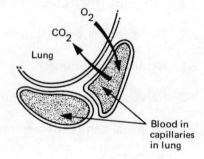

The transfer of gases from one area to another depends entirely on the difference in the pressures of the gases in the different areas. A gas always diffuses from an area where its pressure is high to an area where its pressure is low.

The atmosphere contains oxygen at a pressure of 150 mmHg and almost no carbon dioxide. The tissues contain oxygen at a pressure of 40 mmHg and carbon dioxide at a pressure of 46 mmHg. These pressure differences cause the exchange of gases.

The speed at which the gases are exchanged depends on the extent of the exposure of blood to the air within the lungs.

The respiratory system includes:

the nasal cavities which warm, humidify, and partially cleanse the inspired air;

the larynx (Adam's apple) which is responsible for voice production and for guarding the airway against the entry of food or liquids, since it causes coughing when stimulated;

the trachea which divides into

the bronchi. These branch repeatedly within the lung, finally ending in tiny air sacs — *alveoli.*

The lungs are spongy elastic structures. They entirely fill the thoracic cavities which lie within the rib cage on either side of *the mediastinum.* (The mediastinum is a solid block of structures lying behind the breast bone. It includes the heart, major arteries and veins, oesophagus, and trachea.)

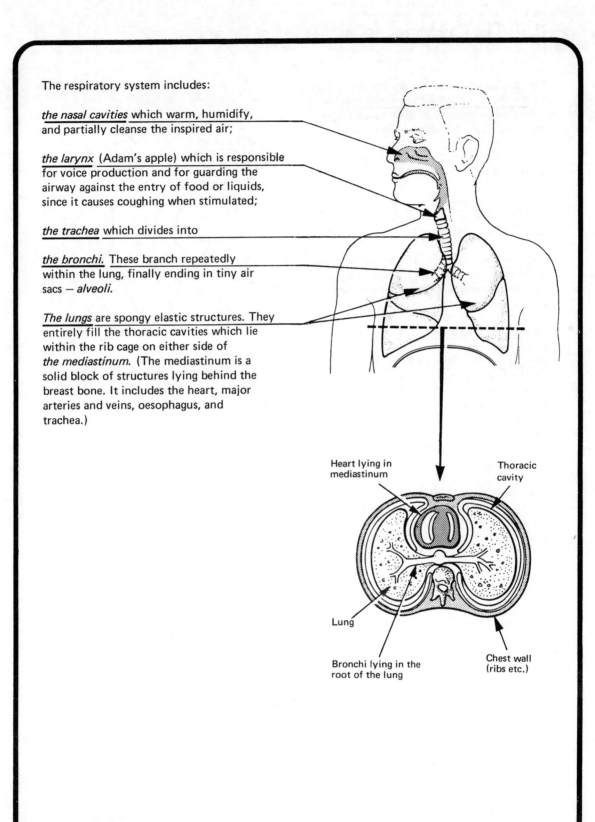

Heart lying in mediastinum

Thoracic cavity

Lung

Bronchi lying in the root of the lung

Chest wall (ribs etc.)

1.2. The air passages

The nasal cavity contains shelf-like projections, *the turbinates,* which act like radiator baffles to warm and moisten the air. The mucosa of the cavity has a rich and variable blood supply.

The mouth is normally used for respiration when extra oxygen is required.

The soft palate can shut off the mouth from the oesophagus and nose and permit respiration during chewing.

The larynx is a complex valve at the junction of the food and air passages. It rides up beneath the tongue during swallowing and so prevents food entering the trachea.

The trachea is kept open by C-shaped rings of cartilage. It divides into a pair of major *bronchi,* which branch further. These larger airways have plates of cartilage in their walls to stop them collapsing during changes in the pressure of the air in the lungs. They are lined with ciliated, mucus-producing epithelium. Dust caught on the mucus is swept up to the larynx by the cilia and coughed out.

The bronchi repeatedly divide and become smaller, forming bronchioles which contain no supporting cartilage, but have walls of smooth muscle which can contract to narrow the airway.

The terminal bronchioles finally open into short, thin-walled passages, respiratory bronchioles, which end in clusters of thin-walled sacs, the *alveoli.*

The total surface area of the alveoli is 50 m^2.

The alveoli are covered in a network of fine capillaries containing blood. Air and blood are in contact across the thin walls, only two cells thick. Here, and only here, exchange of gases by diffusion occurs. (See diagram on page 2.)

A protein called surfactent lowers the surface tension of the fluid film on the alveoli and thus prevents the alveoli collapsing.

Surfactent is produced by the alveolar cells. In its absence the lung tissue would rapidly become solid and airless.

1.3. Gas exchange in the lungs

During quiet respiration about 500 ml of atmospheric air is taken into the lungs with each breath. This air is composed of about 21% oxygen, 79% nitrogen (which plays no part in respiration) and almost no carbon dioxide. The pressure due to the oxygen in the air is about 150 mmHg.

Of the 500 ml inspired:

150 ml fills the mouth trachea and bronchi and takes no part in gas exchange;

350 ml reaches the alveoli and mixes with the "stale" air already present there.

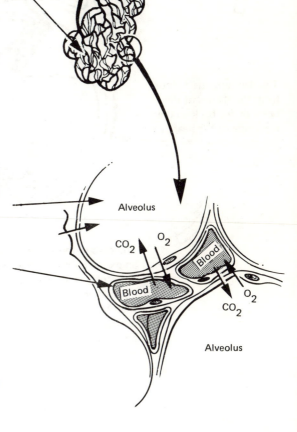

This mixture of inspired and "stale" air has an oxygen pressure of 100 mmHg and a carbon dioxide pressure of 50 mmHg.

The blood in the capillaries covering the alveoli has an oxygen pressure of 40 mmHg and a carbon dioxide pressure of 46 mmHg.

The great difference in pressure causes the oxygen to diffuse rapidly from the alveoli to the blood until its pressure in both is about 100 mmHg.

The difference in carbon dioxide pressures is much less, but this gas is very soluble and diffusion is just as rapid, carbon dioxide passing from the blood to the alveoli until its pressure in both is about 40 mmHg.

Both of these exchanges are completed within milliseconds.

1.4. Gas exchange in the tissues

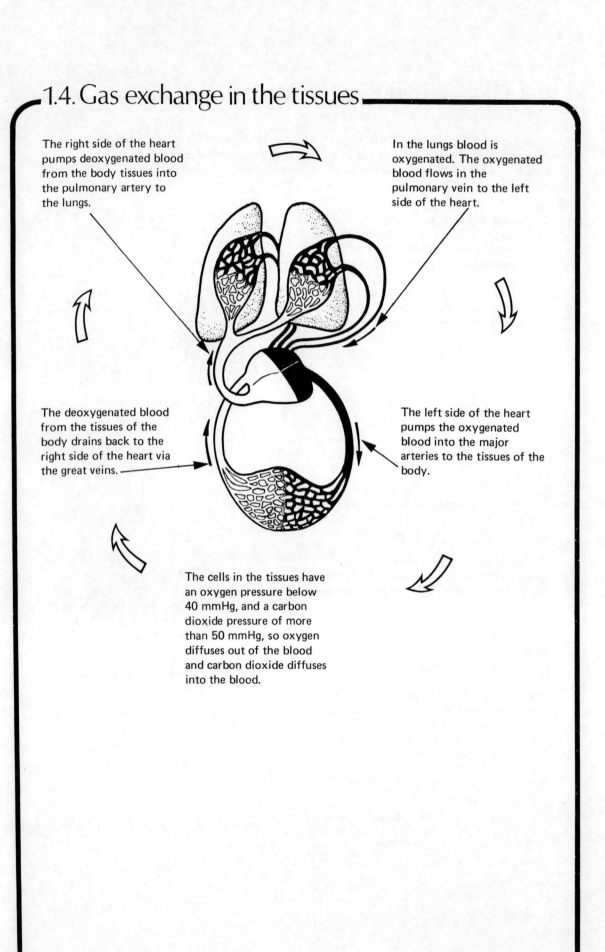

The right side of the heart pumps deoxygenated blood from the body tissues into the pulmonary artery to the lungs.

In the lungs blood is oxygenated. The oxygenated blood flows in the pulmonary vein to the left side of the heart.

The deoxygenated blood from the tissues of the body drains back to the right side of the heart via the great veins.

The left side of the heart pumps the oxygenated blood into the major arteries to the tissues of the body.

The cells in the tissues have an oxygen pressure below 40 mmHg, and a carbon dioxide pressure of more than 50 mmHg, so oxygen diffuses out of the blood and carbon dioxide diffuses into the blood.

1.5. Gas transport

Oxygen Transport

Oxygen is not very soluble in water and not enough could be carried in a simple aqueous solution to keep the tissues alive. But large quantities of oxygen are carried in the blood. The blood contains cells (red corpuscles) packed full of a red pigment known as *haemoglobin*. Haemoglobin is a combination of *haem,* (an iron-porphyrin compound), and *globin,* (a protein). Haemoglobin binds with oxygen to form oxyhaemoglobin (HbO_2) when the gas is present at high pressures. Oxyhaemoglobin releases oxygen at low pressures to form (reduced) haemoglobin (Hb) again.

At oxygen pressures of 100mmHg, as in the alveolar capillaries, all the haemoglobin is oxygenated.

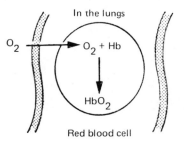

Very little oxygen is released until the oxygen pressure falls below 60mmHg, and most is released at oxygen pressures of 40 mmHg, so the bulk of the oxygen is released in the tissues.
High levels of carbon dioxide and acid (conditions found in active tissues) both increase the release of oxygen.

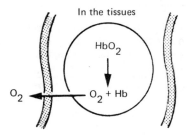

All the haemoglobin is found in the red blood cells. Any free haemoglobin is rapidly excreted by the kidney.

Haemoglobin in babies before birth is different from adult haemoglobin. It is fully oxygenated at lower pressures and therefore carries oxygen more efficiently from the placenta to the baby's circulation.

Carbon Dioxide Transport

In the body tissues, where its concentration is relatively high, carbon dioxide combines with water in the red blood corpuscles to form bicarbonate ions (HCO_3^-) and hydrogen ions (H+). The red blood corpuscles contain an enzyme, carbonic anhydrase, which speeds up this reaction. The bicarbonate ions diffuse out of the corpuscles into the plasma.

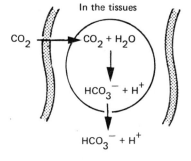

When the bicarbonate ions reach the lungs, where carbon dioxide concentration is relatively low, carbon dioxide and water are re-formed, and the carbon dioxide is released as a gas.

Carbon dioxide is also carried in the blood in solution in the plasma, and combined with protein molecules.

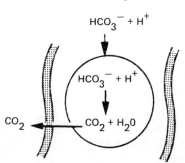

1.6. Voice production

The main function of *the larynx* is to protect the airway, but it is also the organ of voice production.

It is a cartilaginous box, formed in front by the *thyroid cartilage* and behind by the *cricoid cartilage.*

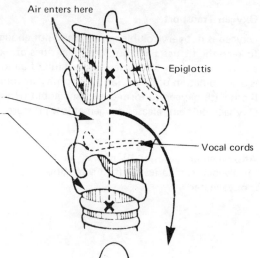

Air enters here

Epiglottis

Vocal cords

The epiglottis lies above like a lid.

Two ligamentous bands pass forward from the top of the cricoid cartilage to the front of the thyroid cartilage. These bands are the *vocal cords* and they act like a pair of shutters. They close the airway when brought together and they open it fully when they move apart. They are controlled by several small muscles.

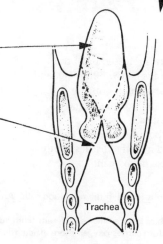

A vertical section through the larynx

Trachea

During breathing the vocal cords lie apart and air moves freely between them.

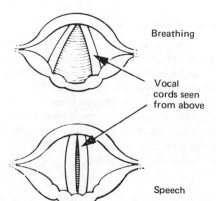

Breathing

Vocal cords seen from above

During voice production they lie close together with only a slit between them. When air is expelled from the lungs the vocal cords vibrate and produce sounds.

Speech

The pitch of the sound depends on the tension of the cords. The quality and volume of the sound is affected by the mouth, nose, sinuses, pharynx, and chest, which act as resonators.

The sounds are converted into speech by the lips, tongue and teeth.

Any object touching the vocal cords causes them to come rapidly together. There is then an involuntary expiratory effort against the closed cords which raises the air pressure in the trachea. The cords spring apart momentarily and a violent gust of air blows the object away. This is *coughing.* During a violent cough the airblow in the trachea can approach the speed of sound.

TEST ONE

1. **Which of the following statements is true?**

   (a)   Respiration is the uptake of oxygen by the cells.

   (b)   Respiration is the exchange of oxygen and carbon dioxide between the cells and the environment.

   (c)   Respiration is the use of energy in the mitochondria to produce carbon dioxide.

2. **Which of the statements on the right apply to the parts of the respiratory system listed on the left?**

i)	The turbinates	(a)	Prevents food entering the airways
ii)	The larynx	(b)	Contain much smooth muscle
iii)	The trachea	(c)	Warm and moisten the air
iv)	The terminal bronchioles	(d)	Supported by C shaped cartilage

3. **What is the pressure of oxygen in the atmosphere?**

   (a) 760 mmHg   (b) 250 mmHg   (c) 150 mmHg   (d) 40 mmHg

4. **Are the following statements true or false?**

		True	False
(a)	There is very little carbon dioxide in the atmosphere.	( )	( )
(b)	All of the haemoglobin of the blood is within the red cells.	( )	( )
(c)	In quiet respiration about 50 ml of air enters the lung.	( )	( )
(d)	Surfactent increases the surface tension in the alveoli.	( )	( )

5. **Which of the following vessels contain deoxygenated blood?**

   (a) Pulmonary artery   (b) Pulmonary vein   (c) A major artery   (d) A great vein.

6. **In which of the following activities do the vocal cords come together?**

   (a) Normal breathing   (b) Voice production   (c) Coughing

ANSWERS TO TEST ONE

1. (b)   Respiration is the exchange of oxygen and carbon dioxide between the cells and the environment.

2.
i)	The turbinates	(c)	warm and moisten the air
ii)	The larynx	(a)	prevents food entering the airways
iii)	The trachea is	(d)	supported by C-shaped cartilage
iv)	The terminal bronchioles	(b)	contain much smooth muscle

3. (c)   150 mmHg

4.

		True	False
(a)	There is very little carbon dioxide in the atmosphere.	(✓)	( )
(b)	All of the haemoglobin of the blood is within the red cells.	(✓)	( )
(c)	In quiet respiration about 50 ml of air enters the lung.	( )	(✓)
(d)	Surfactent increases the surface tension in the alveoli.	( )	(✓)

5. (a) Pulmonary artery   (d) A great vein

6. (b) Voice production   (c) Coughing

2. THE MECHANICS OF RESPIRATION

2.1. The mechanics of respiration

The lungs resemble a balloon sealed within a simple pump, with the neck of the balloon open to the air. As the plunger of the pump is withdrawn, the balloon expands as a partial vacuum is created within the pump. Air is drawn into the balloon, and it expands.

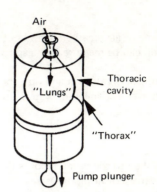

The lungs are contained in a closed cavity — the thoracic cavity — within the thorax. The thoracic cavity is enclosed by:

    the ribs,

    the intercostal muscles,

    the sternum,

    the diaphragm,

    and the vertebral column.

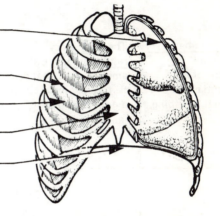

The lungs and the thoracic cavity are lined with a thin moist membrane, the *pleura.* This lines the inside of the thoracic cavity (the *parietal pleura*) and runs over the root of the lung to invest the surface of the lung (the *visceral pleura*) as a continuous layer. It allows the lung to move freely in the thoracic cavity.

Normally the lungs are expanded to completely fill the thoracic cavity. The parietal and visceral pleura lie in contact with one another and the *pleural space* between them is filled by a thin film of lubricant fluid.

The lung is stretched to fill the thorax, as the expanded balloon of the simple model is stretched.

If a hole is made into the chest (e.g. a stab wound) so that air can leak into the thoracic cavity the stretched elastic lung collapses. The accumulation of air within the thoracic cavity is known as a *pneumothorax.*

During inspiration air is drawn into the lungs due to an increase in all the dimensions of the thoracic cage.

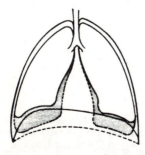

1. The depth of the thorax increases when the diaphragm contracts and descends.

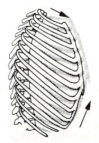

2. The intercostal muscles raise the ribs and increase the fore-and-aft diameter of the thorax, and cause the width of the rib cage to increase.

2.2. Respiratory volumes

The volume of air breathed in and out during respiration can be measured on a spirometer.

As the subject breathes in and out, the drum (inverted over a chamber of water) rises and falls.

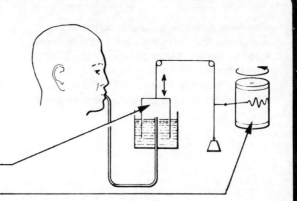

A pen, linked to the drum, records the relative depths of inspirations and expirations on paper on a revolving cylinder.

The tidal volume (about 500 ml) is the amount of air inspired and expired during quiet breathing.

The inspiratory reserve volume is the amount of air that can be forcibly inspired after a normal inspiration (about 2500 ml).

The expiratory reserve volume is the amount of air that can be forcibly expired after a normal expiration (about 1000 ml).

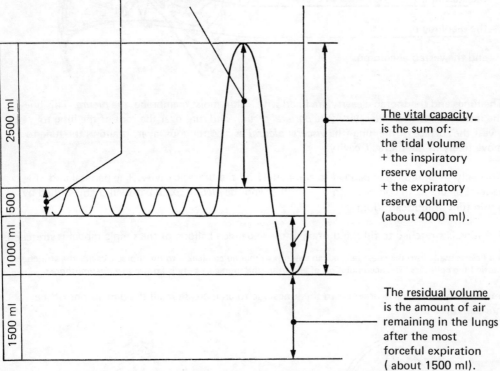

The vital capacity is the sum of: the tidal volume + the inspiratory reserve volume + the expiratory reserve volume (about 4000 ml).

The residual volume is the amount of air remaining in the lungs after the most forceful expiration (about 1500 ml).

All of these volumes may be considerably altered by disease.

Obstruction of the smaller bronchi, (as for example in chronic bronchitis and asthma) causes a slowing of the rate of expiration. This can be measured by determining the most rapid rate of expiration which can be obtained (the peak flow) and the volume which can be expired in one second (the forced expiratory volume). A normal person can expire about 80% of his vital capacity in one second.

Narrowing of large airways, in particular when the larynx is diseased, causes difficulty with inspiration, and breathing is then very noisy — *stridor.*

2.3. Nervous control of respiration

There is no well defined "respiratory nerve centre". Respiration is controlled by nerve cells in the reticular formation of the brain stem, particularly in the medulla. These cells send impulses down the spinal cord and then via the phrenic nerve to the diaphragm, and via the intercostal nerves to the intercostal muscles.

The reticular formation has an inbuilt rhythmical pattern of activity which maintains the rhythmical activity of these muscles.

The rhythm is supplemented by the *Hering-Breuer reflex.* Stretch receptors in the lung tissue send impulses via the vagus nerve to the brain stem. The impulses inhibit inspiration when the lungs are distended, and stimulate inspiration when they are deflated.

In addition pain, and nerve impulses from exercising limbs, cause an increase in the rate and depth of breathing, by their action on the reticular formation.

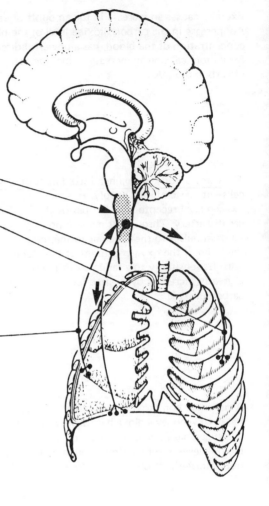

These reflexes, together with the chemical control of respiration considered on the next page, cause the rate and depth of respiration to increase during exercise, and keep the blood gases normal even in heavy exercise.

2.4. Chemical control of respiration

Exercise causes an increase in the amounts of carbon dioxide and acid produced by the muscles. An increase in the carbon dioxide level of the blood, or a rise in the hydrogen ion (H+) concentration of the blood, has a powerful direct effect on the neurones of the reticular formation, causing an increase in the rate and depth of respiration with an increased excretion of carbon dioxide.

The carotid body is a pinhead-sized mass of cells with a rich blood supply. It is an oxygen level receptor and lies beside the carotid artery. Through its nerve supply, it sends impulses to the reticular formation of the brain, if the oxygen content of the blood falls. Its sensitivity to a fall in oxygen level is greatly increased by a raised level of carbon dioxide.

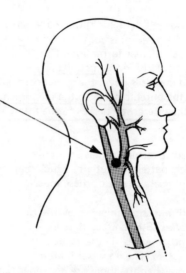

A rise in carbon dioxide level alone is a very powerful stimulus to increased respiration, but a fall in oxygen level alone has surprisingly little effect. A person breathing an atmosphere low in oxygen has none of the sense of suffocation which is promptly felt when the carbon dioxide level is high. He may collapse with oxygen starvation before much increase in respiratory effort is seen, or before it is realised that anything is amiss.

TEST TWO

1. Are the following statements true or false?

		True	False
(a)	During inspiration the diaphragm contracts.	( )	( )
(b)	During inspiration the diaphragm relaxes.	( )	( )
(c)	During inspiration the diaphragm descends.	( )	( )
(d)	During inspiration the diaphragm rises.	( )	( )

2. Is the pressure in the thoracic cavity:

(a) greater than atmospheric pressure?

(b) less than atmospheric pressure?

(c) equal to atmospheric pressure?

3. Which of the statements on the right apply to the quantities listed on the left?

i)	The tidal volume	(a)	Is the air remaining after forceful expiration
ii)	The vital capacity	(b)	Is the maximal volume expired after a forceful inspiration
iii)	The residual volume	(c)	Is the volume inspired during quiet breathing.

4. Are the following statements true or false?

		True	False
(a)	Pure oxygen deprivation causes little sense of suffocation.	( )	( )
(b)	There are several well defined respiratory centres in the brain stem.	( )	( )
(c)	Nervous impulses from exercising limbs increase respiration.	( )	( )
(d)	Inspiration is inhibited by lung distension.	( )	( )

ANSWERS TO TEST TWO

1.

		True	False
(a)	During inspiration the diaphragm contracts.	(✓)	( )
(b)	During inspiration the diaphragm relaxes.	( )	(✓)
(c)	During inspiration the diaphragm descends.	(✓)	( )
(d)	During inspiration the diaphragm rises.	( )	(✓)

2. (b)   less than atmospheric pressure.

3.

i)	The tidal volume	(c)	is the volume inspired during quiet breathing.
ii)	The vital capacity	(b)	is the maximal volume expired after a forceful inspiration
iii)	The residual volume	(a)	is the air remaining after forceful expiration

4.

		True	False
(a)	Pure oxygen deprivation causes little sense of suffocation.	(✓)	( )
(b)	There are several well defined respiratory centres in the brain stem.	( )	(✓)
(c)	Nervous impulses from exercising limbs increase respiration.	(✓)	( )
(d)	Inspiration is inhibited by lung distension.	(✓)	( )

POST TEST

1. **Which of the gas pressures listed on the right are present at the sites listed on the left?**

    i)   In the atmosphere

    ii) In the alveoli

    iii) In venous blood

    (a) Oxygen pressure — 40 mmHg
         Carbon dioxide pressure — 46 mmHg

    (b) Oxygen pressure — 150 mmHg
         Carbon dioxide pressure — about 0 mmHg

    (c) Oxygen pressure — 100 mmHg
         Carbon dioxide pressure — 40 mmHg

2. **Which of the following structures remove dust particles from the bronchi?**

    (a) Goblet cells   (b) Cilia   (c) Cricoids   (d) Alveoli

3 (a) **What is the name of the dome-shaped muscle between the thorax and the abdomen?**

  (b) **What is the name of the block of structures lying between the lungs?**

4. **Are the following statements true or false?**

		True	False
(a)	The right side of the heart pumps deoxygenated blood.	( )	( )
(b)	The terminal bronchioles are kept open by cartilage.	( )	( )
(c)	A rising acid level in the blood increases respiratory effort.	( )	( )
(d)	Oxygenated blood is bright red.	( )	( )

5. **What happens to the lung if air freely enters the thoracic cavity?**

6. **Indicate which of the statements below apply best to asthma and which apply best to laryngeal disease, by placing ticks in the appropriate boxes.**

		Asthma	Laryngeal disease
(a)	Involves obstruction of the small bronchioles.	( )	( )
(b)	Involves obstruction of the larger airways.	( )	( )
(c)	Causes difficulty with inspiration.	( )	( )
(d)	Causes difficulty with expiration.	( )	( )

7. **Which of the statements on the right apply to the items listed on the left?**

    i)   The carotid body

    ii) The cricoid

    iii) The phrenic nerve

    (a) Is connected to the diaphragm

    (b) Is involved in oxygen level detection

    (c) Is involved in voice production

ANSWERS TO POST TEST

1. i) In the atmosphere   (b)   Oxygen pressure — 150 mmHg
                                       Carbon dioxide pressure — about 0 mmHg

     ii) In the alveoli   (c)   Oxygen pressure — 100 mmHg
                                         Carbon dioxide pressure — 40 mmHg

   iii) In venous blood   (a)   Oxygen pressure — 40 mmHg
                                         Carbon dioxide pressure — 46 mmHg

2. (b) Cilia

3. (a) The diaphragm

     (b) The mediastinum

4.

		True	False
(a)	The right side of the heart pumps deoxygenated blood.	(✓)	( )
(b)	The terminal bronchioles are kept open by cartilage.	( )	(✓)
(c)	A rising acid level in the blood increases respiratory effort.	(✓)	( )
(d)	Oxygenated blood is bright red.	(✓)	( )

5. The lung collapses

6.

		Asthma	Laryngeal disease
(a)	Involves obstruction of the small bronchioles.	(✓)	( )
(b)	Involves obstruction of the larger airways.	( )	(✓)
(c)	Causes difficulty with inspiration.	( )	(✓)
(d)	Causes difficulty with expiration.	(✓)	( )

7. i) The carotid body   (b)   is involved in oxygen level detection
     ii) The cricoid   (c)   is involved in voice production
   iii) The phrenic nerve   (a)   is connected to the diaphragm

CONTENTS

THE CARDIOVASCULAR SYSTEM

1. INTRODUCTION

1.1. A general description

The cardiovascular system is the transport system of the body, carrying respiratory gases, foodstuffs, hormones, and other material to and from the body tissues.

The cardiovascular system is made up of:

— the <u>blood</u>, a complex fluid tissue containing specialised cells in a liquid plasma;

— the <u>heart</u>, a double pump containing four chambers, which pumps the blood into the blood vessels;

— the <u>blood vessels</u>:

— <u>arteries</u> which carry blood from the heart to the tissues;

— <u>veins</u> which bring blood back from the tissues to the heart; and

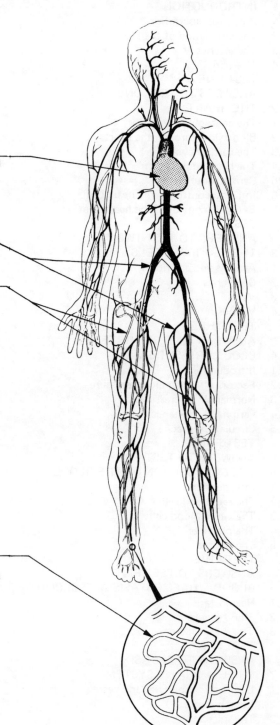

— <u>capillaries</u> which are minute blood vessels found throughout the tissues. They connect the small arteries to the small veins. Exchange of respiratory gases and nutrients with the tissues occurs across the walls of the capillaries.

1.2. The circulation of the blood

The driving force of the circulation is the pressure developed by the contraction of the heart.

The heart is, in fact, two pumps. A large pump — the left heart — pumps blood into the systemic circulation (that is to all the blood vessels of the body apart from those in the lungs). The blood drains from the systemic circulation into the right heart which pumps the blood at a lower pressure to the lungs.

The two parts of the heart form a single anatomical structure but the two blood flows through the heart are separated by a wall — the septum — and do not mix at any point.

A notable exception to the general arrangement in which blood from the systemic circulation drains into veins which empty into the right side of the heart, is that all the blood from the gut first of all flows to the liver, via the portal vein. Only after passing through the liver does the blood empty into the systemic veins.

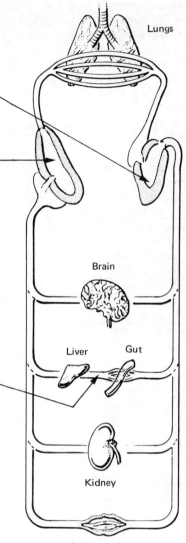

Lungs

Brain

Liver   Gut

Kidney

Skin, muscle, etc.

1.3. The arteries

The main systemic artery is the *aorta*. The aorta is thicker than its owner's thumb. It rises from the heart and then curves over the heart and pulmonary vessels to descend in front of the vertebral column, through the diaphragm into the abdomen.

The main branches of the aorta in the chest are:

the subclavian arteries to the thorax and arm; the carotid arteries to the head, neck and brain; and the coronary arteries to the heart muscle.

The abdominal aorta gives off three arteries to the stomach and gut: the coeliac artery, the superior mesenteric artery, and the inferior mesenteric artery, and the two large, short renal arteries to the kidneys.

The aorta then divides into the common iliac artery, the internal iliac artery, which supplies the pelvic organs, and the external iliac artery which becomes the femoral artery to the legs, and can easily be felt in the groin.

The pulmonary artery has no connection with the other arteries. It carries blood from the right heart to the lungs. It divides at a T-junction soon after leaving the heart. A branch passes to each lung where it branches repeatedly.

Coronary arteries

Each subclavian artery divides into a vertebral artery which supplies the brain stem, and a brachial artery which supplies the arm.

The brachial artery divides just below the elbow into ulnar and radial arteries. The pulsation of the radial artery can be felt on the thumb side at the wrist.

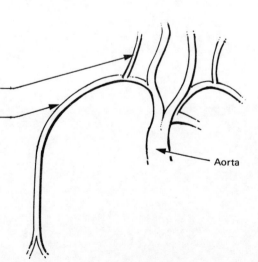

Aorta

1.4. The veins

There are two main systemic veins: the superior vena cava which drains blood from above the diaphragm into the heart; and

the inferior vena cava which drains blood from below the diaphragm.

The four pulmonary veins have no connection with the systemic veins. They empty oxygenated blood directly into the left atrium of the heart.

Blood from the gut, stomach, pancreas and spleen, does not empty directly into the inferior vena cava. It drains into the portal vein and is carried to the liver. After passing through the liver substance the blood is carried via short hepatic veins to the inferior vena cava, which then pierces the diaphragm and enters the right atrium of the heart.

The internal jugular vein drains blood from the brain (via the venous sinuses in the skull) and the head and neck.

The external jugular vein is small.

The subclavian vein from the arms and chest wall joins it to enter the superior vena cava which then runs into the right atrium of the heart.

The hepatic veins, the renal veins, and the common iliac veins are the main tributaries of the inferior vena cava.

In the limbs the deep veins generally follow the arteries. However, there are some large superficial veins just under the skin which are important in temperature regulation.

24

TEST ONE

1. **Are the following statements true or false?**

		True	False
(a)	The heart operates as a single pump.	( )	( )
(b)	The heart operates as a double pump.	( )	( )

2. **Indicate which of the following are functions of arteries, veins or capillaries by placing ticks in the appropriate boxes.**

		Arteries	Veins	Capillaries
(a)	Carry blood to the heart.	( )	( )	( )
(b)	Allow transfer of nutrients and waste products between blood and tissues.	( )	( )	( )
(c)	Carry blood away from the heart.	( )	( )	( )

3. **The diagram on the right represents schematically the pulmonary and systemic circulations.**

   Indicate on the diagram the course taken by de-oxygenated blood like this: – – – – – –➤ and by oxygenated blood like this: ——————➤

   Use arrows to show the direction of flow.

Pulmonary circulation

Heart

Systemic circulation

4. **Indicate the parts of the body supplied by the arteries listed below by placing ticks in the appropriate boxes.**

		Leg	Kidney	Alimentary canal	Head
(a)	Renal artery	( )	( )	( )	( )
(b)	Femoral artery	( )	( )	( )	( )
(c)	Common carotid artery	( )	( )	( )	( )
(d)	Superior mesenteric artery	( )	( )	( )	( )
(e)	Vertebral artery	( )	( )	( )	( )
(f)	Inferior mesenteric artery	( )	( )	( )	( )

5. **What are the names of the two main veins which enter the right atrium of the heart?**

ANSWERS TO TEST ONE

1.   True   False

(a)   The heart operates as a single pump.   ( )   (✓)

(b)   The heart operates as a double pump.   (✓)   ( )

2.   Arteries   Veins   Capillaries

(a)   Carry blood to the heart.   ( )   (✓)   ( )

(b)   Allow transfer of nutrients and waste
products between blood and tissues.   ( )   ( )   (✓)

(c)   Carry blood away from the heart.   (✓)   ( )   ( )

3.

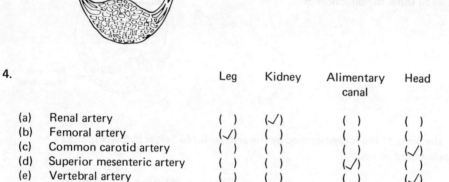

4.

	Leg	Kidney	Alimentary canal	Head
(a) Renal artery	( )	(✓)	( )	( )
(b) Femoral artery	(✓)	( )	( )	( )
(c) Common carotid artery	( )	( )	( )	(✓)
(d) Superior mesenteric artery	( )	( )	(✓)	( )
(e) Vertebral artery	( )	( )	( )	(✓)
(f) Inferior mesenteric artery	( )	( )	(✓)	( )

5. Superior vena cava.
Inferior vena cava.

2. THE HEART

2.1. The anatomy of the heart

The heart is the size of a clenched fist. It lies behind the sternum and costal cartilages in the *mediastinum* — the block of structures between the lungs.

It lies on the central part of the diaphragm in front of the oesophagus. ——

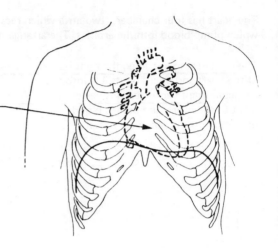

The whole heart lies in the pericardial cavity, a fibrous bag lined with a moist membrane which permits it to move freely during each contraction.

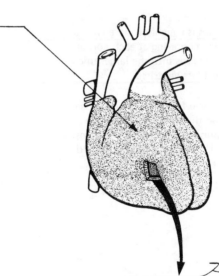

The walls of the heart are composed of branching fibres of cardiac muscle. Under the microscope these fibres can be seen to consist of separate nucleated cells. ——

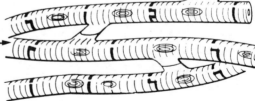

The heart has four chambers, two atria which receive blood from the veins, and two ventricles which pump blood into the arteries. The arrangement of the atria and ventricles is complicated.

The right atrium lies along the right border of the heart and opens on its left into the triangular right ventricle. This lies on the front of the heart, and pumps blood upwards into the pulmonary artery.

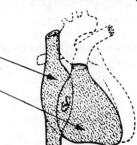

The left atrium is roughly square with a pulmonary vein entering at each corner. It empties downwards into the large, thick-walled conical left ventricle.

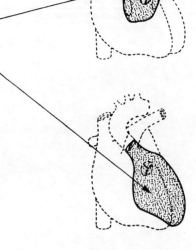

The left ventricle forms the main mass of the heart, and the other chambers are wrapped around it. Its muscle performs the most work, pumping blood upwards into the aorta.

The walls of the atria are thin, but the walls of the ventricles are thick, the left ventricular wall being thicker than the wall of the right ventricle.

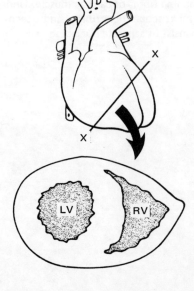

2.2. Blood flow through the heart

Blood flow through the heart is directed by valves which allow the blood to flow in one direction only. Each contraction of the heart simply raises the pressure within the heart chambers. The valves close firmly to prevent any backflow, but float apart to permit free forward flow.

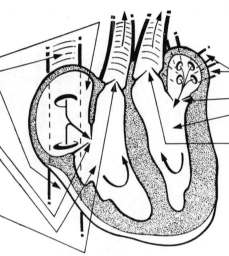

Venous blood from the body tissues enters the right atrium from the superior and inferior venae cavae.

The right atrium pumps the blood through the tricuspid valve into the right ventricle from where it is pumped by contraction of the ventricular wall through semilunar valves into the pulmonary artery on its way to the lungs.

Oxygenated blood from the lungs enters the left atrium through the four pulmonary veins and passes through the mitral valve into the left ventricle from where it is pumped through semilunar valves into the aorta, which distributes the blood to the systemic circulation.

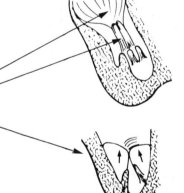

The heart valves are sheets of fibrous tissue. The tricuspid and mitral valves — the *atrioventricular valves* — have to withstand high pressures during the contraction of the heart. Their cusps are attached by chordae tendinae to *papillary muscles* in the wall of each ventricle. When the ventricles contract the cords prevent the valves inverting into the atria.

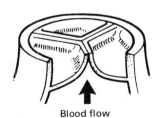

The semilunar valves at the entrances to the aorta and pulmonary artery both consist of three cusps. There are no cords to prevent them inverting since the back pressure to which they are subjected is much less than that exerted on the atrioventricular valves.

Blood flow

A diseased valve which permits backflow is said to be *incompetent*.

A narrowed valve which does not allow free forward flow is said to be *stenosed*.

Mitral stenosis is the commonest valve disease. It is usually caused by scarring from rheumatic fever. By damming the blood back in the lungs it causes breathlessness as its main symptom.

2.3. The cardiac cycle

The heart normally beats about 75 times a minute. The time from the beginning of one heart beat to the beginning of the next (the cardiac cycle) lasts about 0.8 seconds. The sequence of events in the cardiac cycle is outlined in the diagrams below. The two circles at the centre represent the periods of time during which the atria and ventricles are in contraction (systole) and relaxation (diastole).

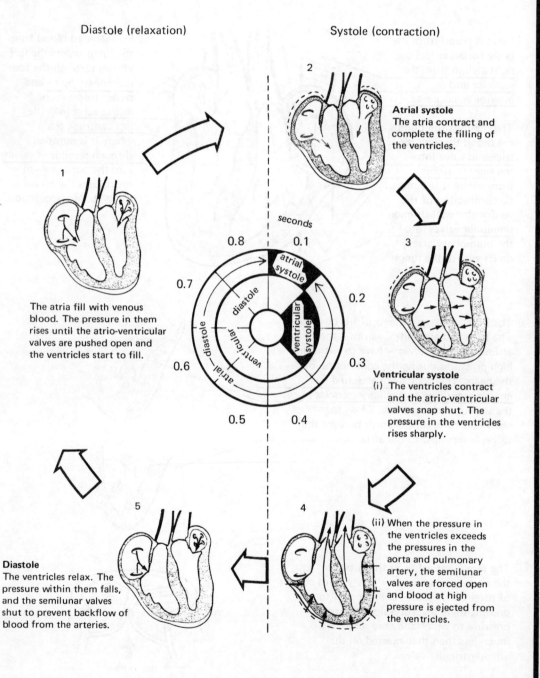

Diastole (relaxation)

Systole (contraction)

2

Atrial systole
The atria contract and complete the filling of the ventricles.

1

The atria fill with venous blood. The pressure in them rises until the atrio-ventricular valves are pushed open and the ventricles start to fill.

seconds

0.8 0.1

0.7

atrial systole

diastole

0.2

0.6   ventricular systole   0.3

atrial diastole

ventricular

3

Ventricular systole
(i) The ventricles contract and the atrio-ventricular valves snap shut. The pressure in the ventricles rises sharply.

0.5   0.4

5

Diastole
The ventricles relax. The pressure within them falls, and the semilunar valves shut to prevent backflow of blood from the arteries.

4

(ii) When the pressure in the ventricles exceeds the pressures in the aorta and pulmonary artery, the semilunar valves are forced open and blood at high pressure is ejected from the ventricles.

2.4. The heart sounds

The blood flow through the heart is practically silent, but when the heart valves close suddenly a thumping sound is heard, similar to that heard when a high pressure tap is suddenly turned off.

Two distinct sounds can be heard during each cardiac cycle.

1. A low pitched sound ("lub") is caused by the sudden closure of the atrioventricular valves when the ventricles begin to contract in early systole. This is the *first heart sound.*

2. A high pitched sound ("dup") is produced by closure of the semilunar valves when the ventricles relax. This is the *second heart sound.*

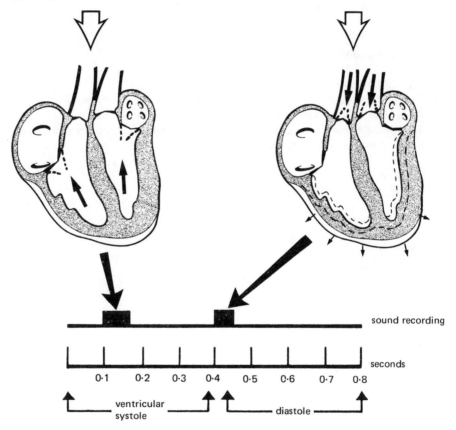

These sounds can readily be heard by placing an ear on the chest, or with a stethoscope. The heart beat (contraction) can be felt with a hand placed on the chest. It occurs between the first and second heart sounds.

If the heart valves are damaged by disease additional sounds (murmurs) will be heard. The nature and timing of murmurs are important in the diagnosis of heart disease. Occasionally murmurs occur in healthy young people.

2.5. Cardiac output

The cardiac output is the amount of blood pumped into the aorta each minute. It is equal to the amount of blood pumped out with each heart beat (the stroke volume) multiplied by the number of beats per minute.

Cardiac output	=	Stroke volume	X	Number of beats/minute

As the stroke volume is normally about 70 ml, and the number of beats per minute is 70—75, the cardiac output is normally about 5 litres per minute.
However, during exercise this may rise to 15 litres or so per minute.

The cardiac output depends on:

(1)   the frequency of the heart beat (see Section 2.7)

(2)   the rate of return of blood to the heart — the *venous return*.
Heart muscle responds to increased stretching by an increase in the force of its contraction. The more it is distended by incoming venous blood, the harder it contracts to expel it. The heart therefore tends automatically to match cardiac output to venous return and thus progressive over-distension of the veins is avoided.

(3)   the blood pressure (see Section 4)

2.6. Intrinsic control of the heart beat

The heart muscle has a nerve supply from the autonomic nervous system but this only serves to modify the heart's action. The heart will continue to beat if its nerve supply is severed. Thus the heart has an intrinsic rhythm of its own.

This rhythm is controlled by a special 'pacemaker' the sino-atrial (SA) node in the wall of the right atrium, from which impulses are continuously discharged.

From the SA node a wave of excitation is conducted across the walls of the atria, causing them to contract.

The atrio-ventricular (AV) node picks up the wave of excitation and relays it by Purkinje fibres in the bundle of His, causing excitation and contraction of the walls of the ventricles.

S.A. node

The excitation wave which spreads through the wall of the heart is accompanied by electrical charges which can be picked up from the chest wall and recorded on an *electrocardiograph.*

The resulting record is called an electrocardiogram (ECG) and its components are:

(1)   the P wave, caused by spread of excitation across the atria

(2)   the QRS wave, caused by spread of excitation through the ventricles

(3)   the T wave, caused by repolarisation (return to resting condition) of the ventricles.

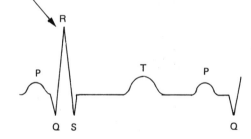

The contraction of heart muscle differs importantly from that of skeletal muscle. There is a long period (about 0.3 sec) after each contraction during which the heart is unable to contract. This *refractory period* prevents the heart undergoing a sustained contraction, which would be lethal.

2.7. Extrinsic control of the heart beat

The intrinsic activity of the heart is controlled and regulated to meet the body's changing needs by impulses originating in the cardiac centres in the medulla oblongata.

Impulses transmitted from these centres down the vagus nerve of the parasympathetic nervous system cause release of acetylcholine at nerve endings in the sino-atrial and atrio-ventricular nodes. These impulses depress the heart's activity by slowing the heart rate and diminishing the strength of contraction.

Conversely, impulses from the cardiac centres travelling down the sympathetic nerves cause release of noradrenaline at the nerve endings, which increases the heart rate and strength of contraction.

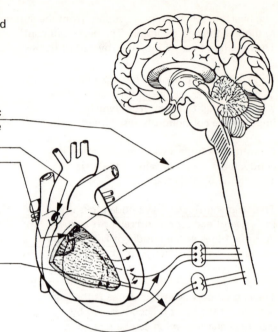

Cardiac Reflexes

Specialised tissues situated in the walls of
- the carotid artery
- the aortic arch
- the right atrium

and nearby parts of the venae cavae respond to changes in blood pressure. Impulses from these pressure receptors reach the cardiac centres and can give rise to an increase or decrease in stimulation along the sympathetic and parasympathetic nerves to restore pressure to normal.

There are also chemoreceptors in the carotid sinus and aortic arch which respond to lack of oxygen, and to an increase in carbon dioxide or H^+ content of the blood.

Any such changes cause a reflex increase in heart rate.

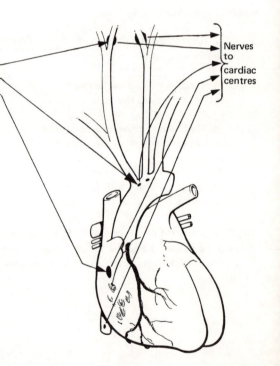

Nerves to cardiac centres

TEST TWO

1. **Indicate which of the following are the functions of the chambers of the heart by placing ticks in the appropriate boxes.**

	Left atrium	Right atrium	Left ventricle	Right ventricle
(a) Receives blood from lungs.	( )	( )	( )	( )
(b) Pumps blood to lungs.	( )	( )	( )	( )
(c) Receives blood from systemic circulation.	( )	( )	( )	( )
(d) Pumps blood to systemic circulation.	( )	( )	( )	( )

2. **Indicate the positions of the valves of the heart by placing ticks in the appropriate boxes.**

	Semilunar valves	Tricuspid valve	Mitral valve
(a) Between left ventricle and left atrium.	( )	( )	( )
(b) Between right ventricle and right atrium.	( )	( )	( )
(c) Between right ventricle and pulmonary artery.	( )	( )	( )
(d) Between left ventricle and aorta.	( )	( )	( )

3. **Mark on the time scale below, by shading the appropriate parts of the bars, the periods of atrial diastole and ventricular diastole.**

Atria
Ventricles
Beginning of atrial systole →
Seconds   0.2   0.4   0.6   0.8

4. **Give three major factors influencing cardiac output.**

5. **Label the tissues indicated on the diagram opposite.**

A
B
C

6. **Are the following statements true or false?**

	True	False
(a) Sympathetic nervous stimulation slows the heart beat.	( )	( )
(b) Sympathetic nervous stimulation accelerates the heart beat.	( )	( )
(c) Parasympathetic nervous stimulation accelerates the heart beat.	( )	( )
(d) Parasympathetic nervous stimulation slows the heart beat.	( )	( )

1.

		Left atrium	Right atrium	Left ventricle	Right ventricle
(a)	Receives blood from lungs.	(✓)	( )	( )	( )
(b)	Pumps blood to lungs.	( )	( )	( )	(✓)
(c)	Receives blood from systemic circulation.	( )	(✓)	( )	( )
(d)	Pumps blood to systemic circulation.	( )	( )	(✓)	( )

2.

		Semilunar valves	Tricuspid valve	Mitral valve
(a)	Between left ventricle and left atrium.	( )	( )	(✓)
(b)	Between right ventricle and right atrium.	( )	(✓)	( )
(c)	Between right ventricle and pulmonary artery.	(✓)	( )	( )
(d)	Between left ventricle and aorta.	(✓)	( )	( )

3.

Beginning of atrial systole →

Atria

Ventricles

Seconds   0.2   0.4   0.6   0.8

4.
(i)   The frequency of the heart beat.
(ii) The venous return.
(iii) The blood pressure.

5. A.   Sino-atrial node.
  B.   Atrio-ventricular node.
  C.   Purkinje fibres.

6.

		True	False
(a)	Sympathetic nervous stimulation slows the heart beat.	( )	(✓)
(b)	Sympathetic nervous stimulation accelerates the heart beat.	(✓)	( )
(c)	Parasympathetic nervous stimulation accelerates the heart beat.	( )	(✓)
(d)	Parasympathetic nervous stimulation slows the heart beat.	(✓)	( )

3. THE BLOOD VESSELS

3.1. Arteries and arterioles

A cross-section of an artery shows that the wall has three layers:

(i)   the tunica intima which consists of a smooth layer of endothelial cells and a band of elastic tissue

(ii) the tunica media which is a mixture of elastic tissue and smooth muscle

(iii) the tunica externa which is fibro-elastic connective tissue.

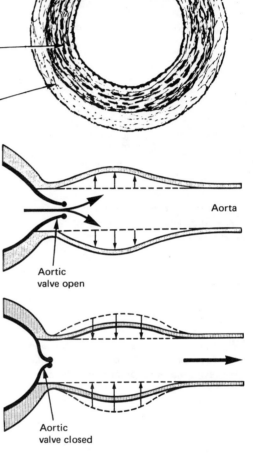

Aorta

Aortic valve open

Aortic valve closed

The aorta and larger arteries not only carry blood to the tissues, but also smooth the flow of blood, by briefly distending with each beat of the heart and then recoiling during diastole. This action converts the intermittent flow of blood from the heart into a steady flow to the tissues. The tunica media of these "pressure-storing" vessels is predominantly elastic.

As the arteries become smaller the proportion of muscular tissue in their walls increases. Vessels with a diameter of less than 0.2 mm are very muscular, and are known as *arterioles*.

Arterioles are thick walled, and have a rich nerve supply. They have a number of very important functions:

1. They act as pressure reducers (rather like the rose on a garden hose-pipe). High arterial blood pressures do not, therefore, reach and damage the tissues.

2. They control local blood flow. If more blood is needed by an organ its arterioles relax and more blood flows through it.

3. They maintain the blood pressure. If all the arterioles were to dilate simultaneously there would be a disastrous fall in blood pressure as the flow of blood to the peripheral circulation would be much greater than the cardiac output. However, due to the overall action of the sympathetic nervous system sufficient arterioles remain constricted to keep the arterial pressure normal.

4. Individual arterioles have a cyclical activity, shutting down and opening up again every few minutes. This ensures that there is a constant turnover of tissue fluid.

3.2. Capillaries

Capillaries are the only vessels in which the blood can perform its function of nutrient/waste exchange. They form a close network which permeates every living tissue (except cartilage and the transparent tissues of the eye).

Venule (small vein)

Arteriole

Capillaries are only one cell thick. They are just large enough (5–10 μm in diameter) to allow red blood cells to pass through them.

The flattened cells which form the walls of capillaries are known as *endothelium.*

Water, electrolytes, and simple molecules can pass readily in and out of the capillaries. However, plasma protein molecules are too large to pass through the endothelium. They remain in the blood and exert an osmotic pressure.

At the arteriolar end of the capillaries the blood pressure is greater than the osmotic pressure of the plasma proteins and fluid filters out of the capillary into the surrounding tissue taking with it oxygen and nutrients.

At the venous end of the capillaries the blood pressure has dropped to a lower level than the osmotic pressure of the plasma proteins and fluid is drawn back bringing with it waste products from the tissues.

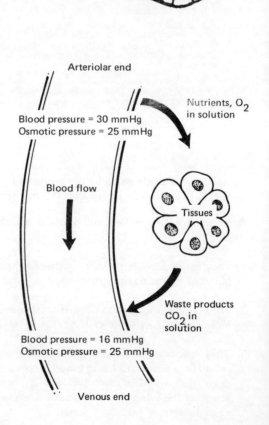

Arteriolar end

Blood pressure = 30 mmHg
Osmotic pressure = 25 mmHg

Nutrients, O_2 in solution

Blood flow

Tissues

Waste products CO_2 in solution

Blood pressure = 16 mmHg
Osmotic pressure = 25 mmHg

Venous end

3.3. Veins

Like arteries, veins have three main layers in their walls. However they are much thinner and distend more easily. They do have smooth muscle in their walls which is under autonomic nervous control.

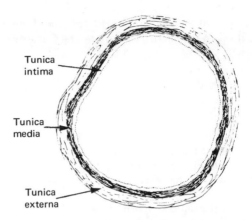

Tunica intima

Tunica media

Tunica externa

Veins over 1 mm in diameter have <u>valves</u> in their walls which direct the flow of blood towards the heart.

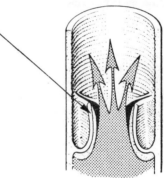

In the limbs there are two sets of veins, the *deep veins* and the *superficial veins.*

The deep veins are often wrapped around the arteries. The blood in these veins, having been cooled during its passage through the limb, tends to absorb heat from the arterial blood and carry it back to the body. This has the effect of reducing heat loss in cold conditions.

The superficial veins (which are very evident in the arms of thin people) are not accompanied by arteries. Heat is readily lost from these veins. The return of venous blood through the superficial veins is increased during warm conditions, thus increasing heat loss from the body.

The veins contain most of the blood in the body.

The great veins are able to alter their capacity by slowly reducing their diameters. This action can compensate for moderate alterations in blood volume over a period of hours, for example after moderate blood loss.

3.4. Venous return

The pressure of the blood in the right atrium of the heart must always be positive, so that blood flows to the heart. There are several factors which ensure the return of venous blood to the heart:

1. *The 'muscle pump'*
   Contraction of muscles, especially those in the legs, squeezes the veins between them and forces blood through the valves towards the heart. ——————

   This mechanism does much of the work of the circulation during exercise.

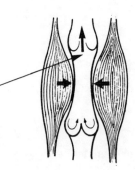

2. *The 'respiratory pump'*
   The negative intrathoracic pressure, which increases during inspiration due to the elastic "pull" of the expanded lungs, tends to suck blood up into the thorax. ——————

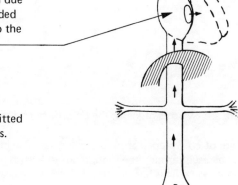

3. *The 'cardiac pump'*
   There is a certain pressure transmitted via the capillaries from the arteries.

4. The pulsations of the arteries tend to milk the veins which lie alongside them.

Effective venous return is vitally important since the heart can only circulate the blood it receives. If the venous return falls, for example after standing still for a long time so that blood pools in the blood vessels of the legs, then the fall in the volume of blood returning to the heart may cause a decrease in the output of the heart sufficient to affect blood flow to the brain and cause fainting.

TEST THREE

1. Indicate the nature of the different parts of the wall of an artery by placing ticks in the appropriate boxes.

		Tunica intima	Tunica media	Tunica externa
(a)	Fibro-elastic connective tissue	( )	( )	( )
(b)	Smooth muscle	( )	( )	( )
(c)	Endothelial cells	( )	( )	( )
(d)	Elastic tissue	( )	( )	( )

2. Which of the following statements is true?

(a) The rate of blood flow is constant from the aorta to the capillaries.
(b) The rate of blood flow decreases from the aorta to the capillaries.
(c) The rate of blood flow increases from the aorta to the capillaries.

3. Name three factors which aid the return of the blood from the veins to the heart.

4. Indicate the parts of the capillaries at which the main exchange of nutrients and waste products takes place by placing ticks in the appropriate boxes.

		Arteriolar end of capillaries	Venous end of capillaries
(a)	Nutrients and oxygen leave the capillaries.	( )	( )
(b)	Waste products and carbon dioxide enter the capillaries.	( )	( )

ANSWERS TO TEST THREE

1.

		Tunica intima	Tunica media	Tunica externa
(a)	Fibro-elastic connective tissue	( )	( )	(✓)
(b)	Smooth muscle	( )	(✓)	( )
(c)	Endothelial cells	(✓)	( )	( )
(d)	Elastic tissue	(✓)	(✓)	( )

2. Statement (b) is true.

3.
(i) The 'muscle pump'
(ii) The 'respiratory pump'
(iii) The 'cardiac pump'

4.

		Arteriolar end of capillaries	Venous end of capillaries
(a)	Nutrients and oxygen leave the capillaries.	(✓)	( )
(b)	Waste products and carbon dioxide enter the capillaries.	( )	(✓)

4. BLOOD PRESSURE

4.1. Introduction

As blood flows through the cardiovascular system it exerts a pressure on the walls of the blood vessels. This pressure is greatest in the arteries close to the heart and lowest in the veins returning blood to the heart.

The pressure of the blood in the arteries varies rhythmically with the beating of the heart reaching a maximum when the left ventricle ejects blood into the aorta (systole) and falling again during diastole reaching a minimum just before the next heart beat.

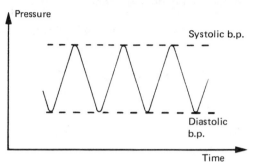

Arterial blood pressure is usually measured with a sphygmomanometer. This consists of an inflatable cuff which is wrapped around the arm and inflated until the brachial artery is occluded (detected by listening through a stethoscope placed over the artery below the cuff).

The pressure in the cuff is measured against a column of mercury and equals the maximum blood pressure, the *systolic blood pressure,* when the artery is just occluded.

The cuff is gradually deflated, allowing blood to spurt through the artery. Tapping sounds are heard through the stethoscope. The noise fades when the flow of blood through the artery is completely uninterrupted. The pressure in the cuff then corresponds to the lowest pressure during the cardiac cycle — the *diastolic blood pressure.*

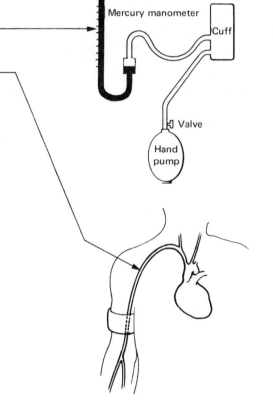

Blood pressure is expressed as two numbers, e.g. 120/80, where 120 represents the systolic blood pressure in millimetres of mercury (mmHg), and 80 represents the diastolic blood pressure in millimetres of mercury.

Normal blood pressure is a variable quantity. A typical healthy young adult may have a blood pressure from about 100/60 to about 150/90. It varies considerably with sleep, exercise and emotion, and tends to rise gradually with increasing age.

4.2. Factors affecting blood pressure

When a liquid flows in a tube, the rate of
flow of the liquid is related to the pressure
exerted on it by the expression:

Pressure   =   Flow rate   X   Resistance to flow.

The blood pressure therefore depends on the
rate of flow of blood through the
circulation, and on the resistance offered to
its flow.

The rate of flow of blood depends on:
a) the blood volume, and
b) the cardiac output.

The resistance to the flow of blood depends on:

c) the viscosity of the blood, and
d) the properties of the blood vessels. The
   blood vessels which offer the greatest
   resistance to the flow of blood are the
   small arteries and arterioles. This
   resistance to blood flow is termed the
   "peripheral resistance".

Blood volume and blood viscosity are
constant in normal healthy individuals, but
cardiac output and peripheral resistance may
vary.

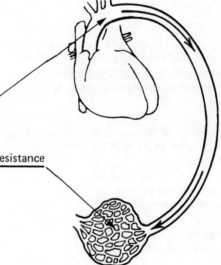

The blood pressure is therefore given by the
expression:

Blood pressure   =   Cardiac output   X   Peripheral resistance

Cardiac output and peripheral resistance
vary widely from individual to individual, and
in one individual at different times. The
blood pressure usually varies to a lesser
extent because the changes in cardiac
output and peripheral resistance tend to
offset one another.

4.3. Normal blood pressures

The diagram below shows typical values for the systolic blood pressure in a resting adult.

In the pulmonary circulation the arterioles are not active and the pressure is low. This avoids the filtration of fluid which occurs in the systemic capillaries and prevents the alveoli from flooding with tissue fluid.

In the systemic circulation the arterioles as a whole are partially contracted. This keeps the peripheral resistance, and hence the blood pressure, high.

Pressure in the right atrium must always be positive, or the heart would not fill. The true pressure here, the *central venous pressure,* can be measured by passing a fine tube along an arm vein, via a subclavian vein into the right atrium. The central venous pressure is quite stable in healthy people due to the continuous adjustment of the venous return.

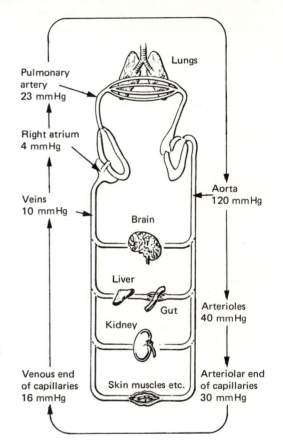

The flow of blood in the arteries is pulsatile and rapid. In the capillaries the flow is slow and smooth.

4.4. Peripheral resistance

Resistance to the flow of fluid through a tube is mainly dependent on the diameter of the tube. Doubling the diameter increases the rate of flow 16 times, if pressure remains constant. Thus minor changes in arteriolar constriction have a great effect on blood flow.

Differential contraction of arterioles supplying various parts of the body acts to shunt blood to the areas where it is most needed.

Thus during digestion the flow of blood to the alimentary canal is increased, and that to the skeletal muscles decreased.

Regulation of 'vasomotor tone' — the degree of contraction of muscular arteries and arterioles — is mainly under the control of a <u>vasomotor centre</u> in the medulla.

This centre influences the activities of the sympathetic and parasympathetic nerves to the various blood vessels.

Parasympathetic (cholinergic) stimulation causes:

vasodilation of arterioles of salivary glands,

vasodilation of arterioles of external genitals.

Sympathetic (adrenergic) stimulation causes:

vasoconstriction of arterioles in skin,

vasoconstriction of arterioles in viscera,

vasodilation of coronary arterioles,

vasodilation of arterioles in skeletal muscle.

(This, in response to sudden fear, can cause a sudden drop in blood pressure, and is the mechanism of *fainting*.)

The sympathetic nervous system maintains a constant *tone* (degree of contraction of the arterioles), which overall is vasoconstrictor and is the main factor in the minute-to-minute maintenance of normal blood pressure.

Control of the vasomotor centre

The main factors influencing the vasomotor centre are shown opposite.

Factors which increase the constant stream of vasoconstrictor impulses to the abdominal and skin arterioles are:

— a fall in blood pressure as measured by pressure receptors in the aortic arch and carotid artery

— a fall in blood oxygen level or a rise in blood carbon dioxide level as measured by chemoreceptors in the aortic arch and carotid artery

— an increase in blood carbon dioxide or a fall in blood temperature acting directly on the centre

— spread of activity from the respiratory centre at inspiration.

All these stimuli initiate *vasoconstrictor reflexes,* which tend to increase blood pressure.

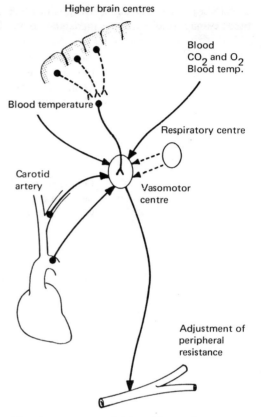

Conversely, *vasodilator reflexes,* acting to decrease blood pressure, are initiated from the vasomotor centre by:

— a rise in blood pressure as measured from the aortic arch and carotid artery
— a fall in blood carbon dioxide or rise in blood oxygen levels.

These and other factors lead to a decrease in the rate of vasoconstrictor impulses sent out by the centre.

Local chemical control of blood vessels

Arterioles can also be influenced to contract or relax by local factors

vasodilation is caused by:	*vasoconstriction* is caused by:
— excess CO_2 — lack of O_2	— adrenaline and noradrenaline — histamine (released e.g. during tissue damage) — other hormones

4.5. Circulatory shock

Circulatory shock is an abnormal state in which the cardiac output is so much reduced that tissue damage results from an inadequate blood flow.

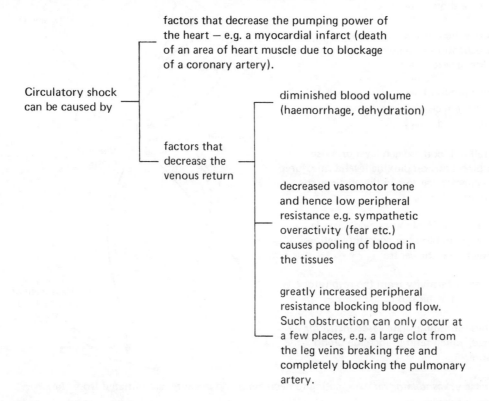

Circulatory shock can be caused by

factors that decrease the pumping power of the heart — e.g. a myocardial infarct (death of an area of heart muscle due to blockage of a coronary artery).

factors that decrease the venous return

diminished blood volume (haemorrhage, dehydration)

decreased vasomotor tone and hence low peripheral resistance e.g. sympathetic overactivity (fear etc.) causes pooling of blood in the tissues

greatly increased peripheral resistance blocking blood flow. Such obstruction can only occur at a few places, e.g. a large clot from the leg veins breaking free and completely blocking the pulmonary artery.

The role of aldosterone in circulatory shock

When circulatory shock is due to reduced blood volume there is an increase in aldosterone output which acts to return the cardiac output and blood volume towards normal values. The mechanism of operation is shown below:

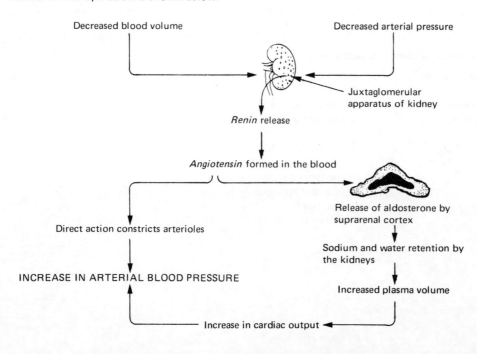

Decreased blood volume

Decreased arterial pressure

Juxtaglomerular apparatus of kidney

Renin release

Angiotensin formed in the blood

Direct action constricts arterioles

Release of aldosterone by suprarenal cortex

Sodium and water retention by the kidneys

INCREASE IN ARTERIAL BLOOD PRESSURE

Increased plasma volume

Increase in cardiac output

TEST FOUR

1. **Blood pressure can be expressed as two numbers, for example: 120/80.**
   In this case:
   (a)   what does 120 stand for?
   (b)   what does 80 stand for?

2. **Define briefly the term "peripheral resistance".**

3. **Which of the blood pressures on the right occur in the blood vessels listed on the left?**

(i)	Aorta	(a)	16 mmHg
(ii)	Pulmonary artery	(b)	10 mmHg
(iii)	Venous end of the systemic capillaries	(c)	23 mmHg
(iv)	Vein	(d)	120 mmHg

4. **Indicate which of the following are effects of sympathetic nervous stimulation and which are effects of parasympathetic nervous stimulation by placing ticks in the appropriate boxes.**

		Sympathetic stimulation	Parasympathetic stimulation
(a)	Vasodilation of coronary arterioles	( )	( )
(b)	Vasoconstriction of skin arterioles	( )	( )
(c)	Vasodilation of skeletal muscle arterioles	( )	( )
(d)	Vasodilation of arterioles of genitals	( )	( )
(e)	Vasoconstriction of visceral arterioles	( )	( )

5. **Give four factors which tend to cause increased blood pressure by increasing the rate of emission of vasoconstrictor impulses from the vasomotor centre.**

ANSWERS TO TEST FOUR

1. (a) The systolic blood pressure (in mmHg)
   (b) The diastolic blood pressure (in mmHg)

2. Peripheral resistance is the resistance offered to the flow of blood by the small blood vessels of the systemic circulation.

3. (i) Aorta   (d) 120 mmHg
   (ii) Pulmonary artery   (c) 23 mmHg
   (iii) Venous end of the systemic capillaries   (a) 16 mmHg
   (iv) Vein   (b) 10 mmHg

4.

		Sympathetic stimulation	Parasympathetic stimulation
(a)	Vasodilation of coronary arterioles	(✓)	( )
(b)	Vasoconstriction of skin arterioles	(✓)	( )
(c)	Vasodilation of skeletal muscle arterioles	(✓)	( )
(d)	Vasodilation of arterioles of genitals	( )	(✓)
(e)	Vasoconstriction of visceral arterioles	(✓)	( )

5. (a) A fall in blood pressure.
   (b) A fall in blood oxygen level.
   (c) A fall in blood temperature.
   (d) An increase in blood carbon dioxide level.

5 THE COMPOSITION OF THE BLOOD

5.1. Introduction

The normal blood volume of an adult is about 5 litres. The blood consists of a suspension of *cells* in a clear fluid *plasma*. If the cells are allowed to settle by spinning in a centrifuge they are found to occupy 40–50% of the total blood volume.

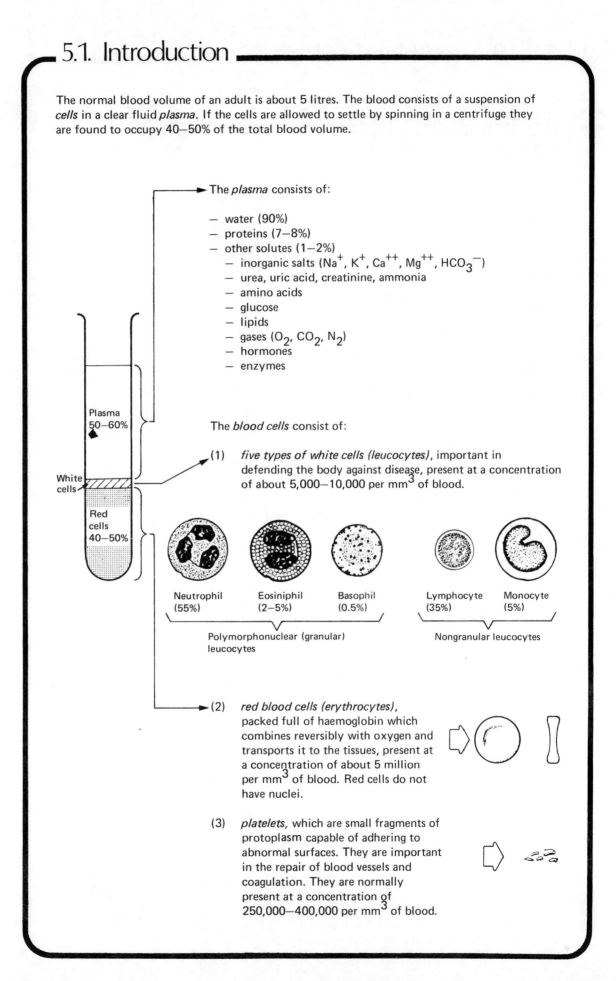

The *plasma* consists of:

— water (90%)
— proteins (7–8%)
— other solutes (1–2%)
  — inorganic salts (Na^+, K^+, Ca^{++}, Mg^{++}, HCO_3^-)
  — urea, uric acid, creatinine, ammonia
  — amino acids
  — glucose
  — lipids
  — gases (O_2, CO_2, N_2)
  — hormones
  — enzymes

The *blood cells* consist of:

(1) *five types of white cells (leucocytes)*, important in defending the body against disease, present at a concentration of about 5,000–10,000 per mm^3 of blood.

| Neutrophil (55%) | Eosiniphil (2–5%) | Basophil (0.5%) | Lymphocyte (35%) | Monocyte (5%) |

Polymorphonuclear (granular) leucocytes   Nongranular leucocytes

(2) *red blood cells (erythrocytes)*, packed full of haemoglobin which combines reversibly with oxygen and transports it to the tissues, present at a concentration of about 5 million per mm^3 of blood. Red cells do not have nuclei.

(3) *platelets,* which are small fragments of protoplasm capable of adhering to abnormal surfaces. They are important in the repair of blood vessels and coagulation. They are normally present at a concentration of 250,000–400,000 per mm^3 of blood.

Plasma 50–60%

White cells

Red cells 40–50%

5.2. The red blood cells

All blood cells develop from primitive reticulum cells. In adults those cells from which the red blood cells, or erythrocytes, are formed are found only in the red marrow of flat bones and the ends of long bones.

The stages in erythropoiesis (the formation of erythrocytes or red blood cells) are shown opposite.

The rate of erythropoiesis is controlled by a glucoprotein — erythropoietin — produced in the liver and kidneys. It regulates the speed of mitosis of reticulum cells into proerythroblasts.

Factors essential for normal erythropoiesis include:
— iron (2mg/day needed),
— vitamin B_{12},
— folic acid,
— vitamin C,
    copper,
    thyroid hormone, corticosteroids.
The process of maturation involves:
— a progressive reduction in cell size,
— a gradual accumulation of haemoglobin,
— the loss of the nucleus.

The number of red cells in the blood remains fairly constant:
— about 5.5 million/mm^3 in men
— about 4.8 million/mm^3 in women.

Erythrocyte production is increased by any conditions which lead to decreased oxygen levels in the tissues (e.g. high altitudes), and following, for example, haemorrhage.

Destruction of red blood cells

After about 120 days, red blood cells 'wear out' and are destroyed by phagocytic endothelial cells in the reticulo-endothelial system of the spleen, liver, and bone marrow.

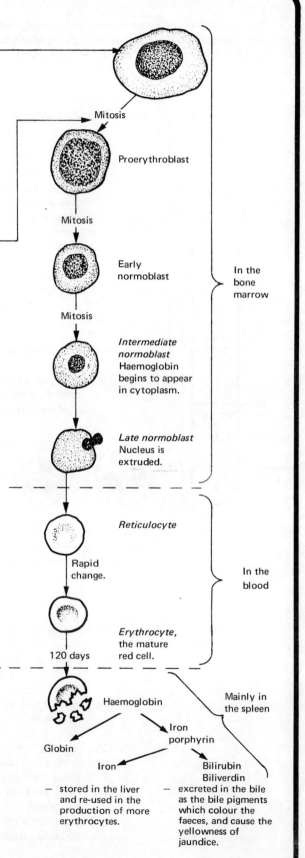

Mitosis

Proerythroblast

Mitosis

Early normoblast

Mitosis

Intermediate normoblast
Haemoglobin begins to appear in cytoplasm.

Late normoblast
Nucleus is extruded.

In the bone marrow

Reticulocyte

Rapid change.

In the blood

Erythrocyte, the mature red cell.

120 days

Haemoglobin

Mainly in the spleen

Iron porphyrin

Globin

Iron

Bilirubin
Biliverdin

— stored in the liver and re-used in the production of more erythrocytes.

— excreted in the bile as the bile pigments which colour the faeces, and cause the yellowness of jaundice.

5.3. The white blood cells

The white blood cells, or leucocytes, are much less numerous in the blood than erythrocytes. They can migrate freely from the blood into the tissues and back by amoeboid movement. They are formed in the bone marrow, the lymph nodes, and the spleen, from reticulum cells:

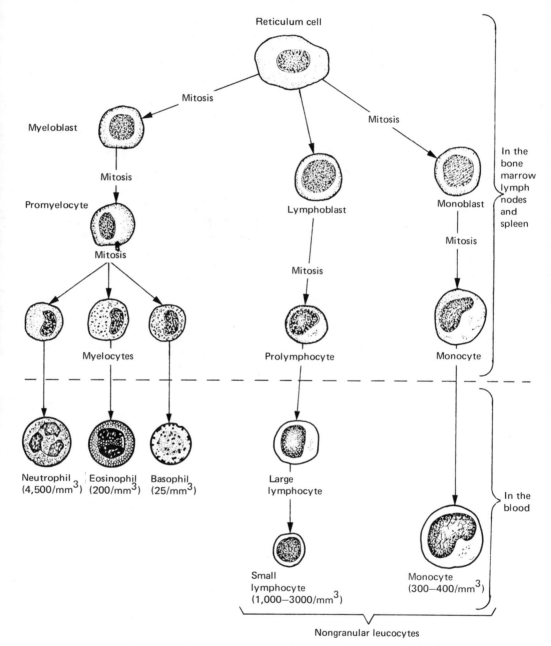

The average total number of leucocytes in the circulation fluctuates between $5,000-10,000/mm^3$ of blood. Exudates and protein fractions from damaged and infected tissues can cause an increase in the rate of production of leucocytes.

Small lymphocytes are particularly active in destroying foreign cells as they possess the property of recognising foreign protein and attacking it directly in the *cell-mediated immune response.*

The life span of leucocytes in the blood is probably about 13–20 days.

5.4. The platelets

Platelets are formed from <u>reticulum cells</u> in the bone marrow.

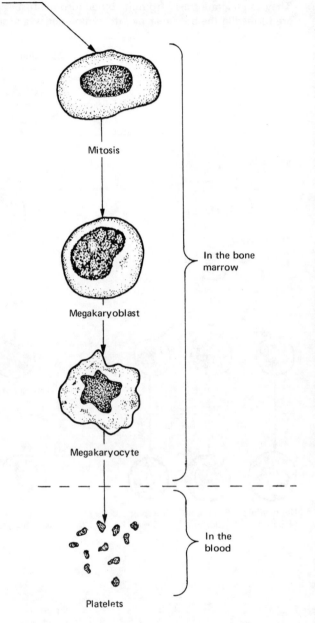

Mitosis

Megakaryoblast

Megakaryocyte

In the bone marrow

Platelets

In the blood

There are about 250,000–500,000 platelets per mm^3 of blood. They tend to stick to damaged surfaces, and there release substances necessary for the coagulation of blood. Platelets last for about 4 days in the blood. They are removed by phagocytic cells in the spleen.

Normally minute breaches in the capillary endothelium are quickly sealed by the action of platelets. However, if the concentration of platelets falls below about 40,000 per mm^3, (as may occur in an allergic reaction to a drug), capillary bleeding readily occurs in the skin, gut and brain.

5.5. The plasma

The plasma makes up about 55% of the volume of the blood.

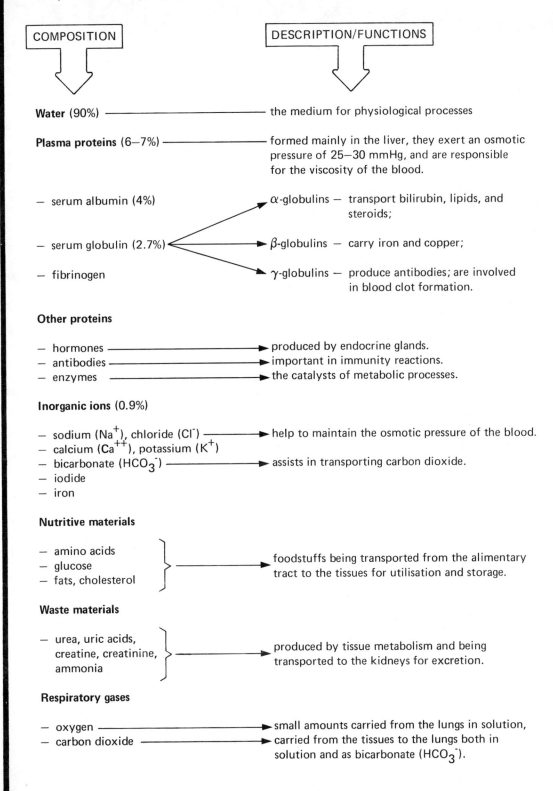

COMPOSITION	DESCRIPTION/FUNCTIONS

Water (90%) ——— the medium for physiological processes

Plasma proteins (6—7%) ——— formed mainly in the liver, they exert an osmotic pressure of 25—30 mmHg, and are responsible for the viscosity of the blood.

— serum albumin (4%)

— serum globulin (2.7%)

— fibrinogen

α-globulins — transport bilirubin, lipids, and steroids;

β-globulins — carry iron and copper;

γ-globulins — produce antibodies; are involved in blood clot formation.

Other proteins

— hormones ——— produced by endocrine glands.
— antibodies ——— important in immunity reactions.
— enzymes ——— the catalysts of metabolic processes.

Inorganic ions (0.9%)

— sodium (Na^+), chloride (Cl^-) ——— help to maintain the osmotic pressure of the blood.
— calcium (Ca^{++}), potassium (K^+)
— bicarbonate (HCO_3^-) ——— assists in transporting carbon dioxide.
— iodide
— iron

Nutritive materials

— amino acids
— glucose
— fats, cholesterol

foodstuffs being transported from the alimentary tract to the tissues for utilisation and storage.

Waste materials

— urea, uric acids, creatine, creatinine, ammonia

produced by tissue metabolism and being transported to the kidneys for excretion.

Respiratory gases

— oxygen ——— small amounts carried from the lungs in solution,
— carbon dioxide ——— carried from the tissues to the lungs both in solution and as bicarbonate (HCO_3^-).

TEST FIVE

1. (a) What is the average number of red cells per mm^3 of blood?
   (b) What is the average number of white cells per mm^3 of blood?

2. Indicate which of the statements below apply to white blood cells and which to red blood cells by placing ticks in the appropriate boxes.

		White cells	Red cells
(a)	Precursor cells are reticulum cells.	( )	( )
(b)	Formed in bone marrow.	( )	( )
(c)	Formed in spleen.	( )	( )
(d)	Formed in lymph nodes.	( )	( )

3. Which type of leucocyte is most active in destroying foreign cells?

4. Which of the statements on the right apply to the plasma proteins listed on the left?

   (i) α-globulin   (a) Involved in blood clotting.
   (ii) β-globulin   (b) Involved in the transport of bilirubin, lipids and steroids.
   (iii) γ-globulin   (c) Involved in the transport of iron and copper.
   (iv) Fibrinogen   (d) Involved in the production of antibodies.

5. Indicate the life spans of the blood cells below by placing ticks in the appropriate boxes.

		∼4 days	13–20 days	∼120 days
(a)	Red blood cells	( )	( )	( )
(b)	Leucocytes	( )	( )	( )
(c)	Platelets	( )	( )	( )

ANSWERS TO TEST FIVE

1. (a) About 5 million.
   (b) About 5—10,000.

2.

		White cells	Red cells
(a)	Precursor cells are reticulum cells.	(✓)	(✓)
(b)	Formed in bone marrow.	(✓)	(✓)
(c)	Formed in spleen.	(✓)	( )
(d)	Formed in lymph nodes.	(✓)	( )

3. Small lymphocyte.

4. (i) α-globulin is (b) involved in the transport of bilirubin, lipids and steroids.
   (ii) β-globulin is (c) involved in the transport of iron and copper.
   (iii) γ-globulin is (d) involved in the production of antibodies.
   (iv) Fibrinogen is (a) involved in blood clotting.

5.

		~4 days	13—20 days	~120 days
(a)	Red blood cells	( )	( )	(✓)
(b)	Leucocytes	( )	(✓)	( )
(c)	Platelets	(✓)	( )	( )

6.1. Haemostasis

Haemostasis is the arrest of bleeding after blood vessel injury. It is partly due to changes in the plasma, partly due to the activity of platelets and partly due to changes in the wall of the injured blood vessel.

When a blood vessel is damaged platelets cover the injured endothelial lining and form a plug which restricts the flow of blood. In addition they release vasoconstrictors, such as serotonin, which cause the vessel to constrict.

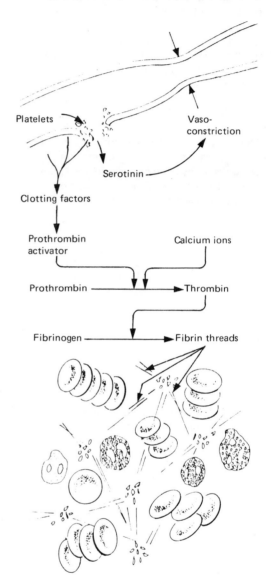

The platelets, the injured endothelium and the local injured tissue, all release *clotting factors* which react with one another, and with soluble circulating proteins to form *prothrombin activator.*

In the presence of calcium ions prothrombin activator converts *prothrombin,* a circulating α-globulin, into *thrombin.*

Thrombin converts *fibrinogen,* a protein present in high concentrations in the blood, into insoluble threads of *fibrin,* which form a solid clot, enmeshing platelets and blood cells.

The clot later shrinks as *serum* (plasma minus fibrinogen) is exuded from it. After several days the clot is destroyed by the enzyme plasmin which digests fibrin.

The complex chain of events involved in the clotting process helps to ensure that clotting does not occur spontaneously in the circulation, which would be disastrous as the supply of food and oxygen to vital organs would be cut off.

6.2. Blood groups

Many foreign substances, especially proteins and polysaccharides, when introduced into the body will stimulate the production of certain γ-globulin proteins called antibodies.

Foreign substances which stimulate the production of antibodies are called antigens. The antibody produced reacts specifically with the antigen which stimulated its production, (otherwise it might attack the body's own proteins).

This reaction is very important in defending the body against bacteria. Proteins on the cell walls of invading bacteria are recognised. Antibodies are formed and the union of antigen with antibody leads to bacterial cell destruction.

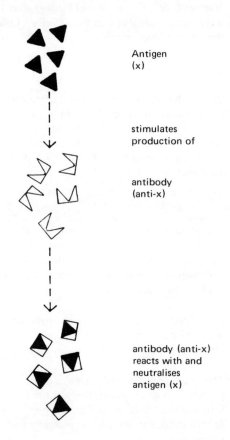

Antigen (x)

stimulates production of

antibody (anti-x)

antibody (anti-x) reacts with and neutralises antigen (x)

Normally, antibodies are only present in the blood of an individual if he has been exposed to the antigen, and they form valuable evidence of previous infection as well as protecting the individual against further attacks.

However, there is one important exception to this rule. Antigens and antibodies related to different blood groups are present in everybody's blood.

An individual's red blood cell can carry many substances on its outer cell membrane. Among these are two different polysaccharides called A and B. An individual may have either, none or both of these, but if he lacks either one his plasma always contains powerful antibodies against the missing factor. The reason for this is unknown.

Blood can be grouped into four types depending on the antigens in the red blood cells and the antibodies in the plasma. These groups are:

Group O (46% of population)	No polysaccharide in cells	Antibodies anti-A and anti-B in plasma
Group A (42% of population)	Polysaccharide A in cells	Antibody anti-B in plasma
Group B (9% of population)	Polysaccharide B in cells	Antibody anti-A in plasma
Group AB (3% of population)	Polysaccharides A and B in cells	No antibodies in plasma

In giving blood transfusions, it is important that the recipient's plasma should not contain antibodies (agglutinins) that will react with the antigen (agglutinogen) in the donor's red cells.

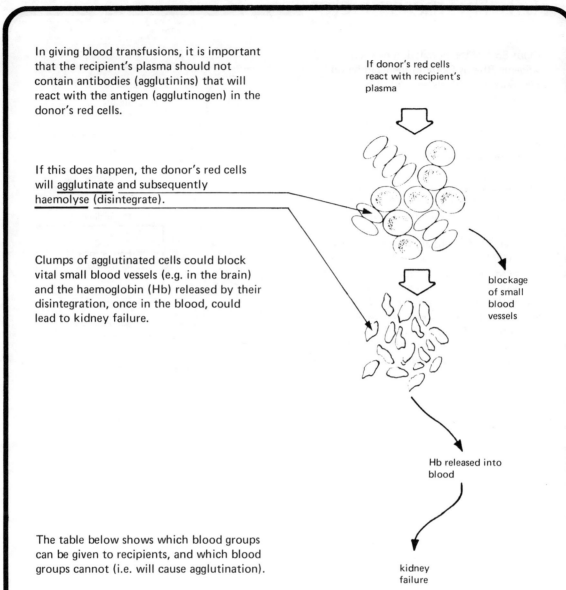

If donor's red cells react with recipient's plasma

If this does happen, the donor's red cells will <u>agglutinate</u> and subsequently <u>haemolyse</u> (disintegrate).

blockage of small blood vessels

Clumps of agglutinated cells could block vital small blood vessels (e.g. in the brain) and the haemoglobin (Hb) released by their disintegration, once in the blood, could lead to kidney failure.

Hb released into blood

kidney failure

The table below shows which blood groups can be given to recipients, and which blood groups cannot (i.e. will cause agglutination).

RECIPIENT			CANNOT RECEIVE (i.e. will cause agglutination)	CAN RECEIVE
Group	Agglutinogen in red cells	Agglutinin in plasma		
O	none	anti-A anti-B	group A group B group AB	group O only
A	A	anti-B	group B group AB	group A group O
B	B	anti-A	group A group AB	group B group O
AB	AB	none		all groups

6.3. The rhesus factor

About 85% of the population have an antigenic Rhesus factor in their red blood cells. Such individuals are said to be Rhesus positive (Rh^+).

The remaining 15% of the population do not posses Rhesus antigens and are said to be Rhesus negative (Rh^-).

There are no naturally occurring Rhesus antibodies corresponding to those in the plasma belonging to the ABO groups.

However, if Rh^+ cells are transfused into a Rh^- individual, antibodies to the Rh^+ cells will be formed in the plasma of the Rh^- recipient. The antibodies are formed relatively slowly, and do not affect the Rh^+ cells of the first transfusion. If a second transfusion of Rh^+ cells is given to the same Rh^- individual, then the antibodies produced previously will cause agglutination and haemolysis of the donor cells.

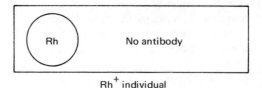

Rh⁺ individual

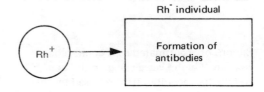

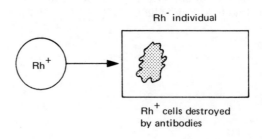

Rh^+ cells destroyed by antibodies

If a Rh^- mother carries a Rh^+ fetus, (which can happen if the father is Rh^+), fetal red cells may cross the placenta into the mother's blood stream and stimulate the production of antibodies. If these antibodies then reach the fetal circulation, serious damage to the fetal red cells may occur

Generally, little harm is done during the first such pregnancy, but second and third Rh^+ fetuses run a serious risk of suffering extensive haemolysis of their red cells. This is called haemolytic disease of the newborn.

These babies may be stillborn, or become deeply jaundiced after birth with the development of severe brain damage. It is therefore extremely important to give young girls only the correct Rh group blood, as a mistake may render them unable to bear living children.

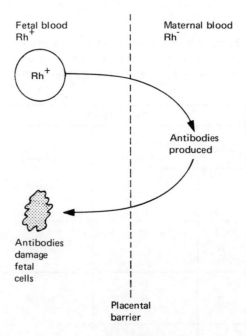

TEST SIX

1. **Which of the statements on the right apply to the substances listed on the left?**

   (i)   Serotonin   (a)   Is converted by thrombin into fibrin threads.
   (ii) Prothrombin   (b)   Is converted into thrombin.
   (iii) Fibrinogen   (c)   Causes vasoconstriction.

2. **What is an antigen?**

3. **What agglutinins, if any, are present in the plasma of individuals in the following blood groups:**

   (a)   Group A?
   (b)   Group O?
   (c)   Group AB?

4. **Which blood groups can be transfused to an individual in the following blood groups:**

   (a)   Group O?
   (b)   Group B?
   (c)   Group AB?

5. **If a Rh^- mother carries a Rh^+ fetus, what may happen to the red blood cells of the fetus?**

1. (i)   Serotonin   (c)   causes vasoconstriction.
   (ii)   Prothrombin   (b)   is converted into thrombin.
   (iii) Fibrinogen   (a)   is converted by thrombin into fibrin threads.

2. An antigen is a foreign substance (usually a protein or polysaccharide) which enters the body and stimulates the production of antibodies.

3. (a)   Anti-B
   (b)   Anti-A, anti-B
   (c)   None

4. (a)   Group O only
   (b)   Group B, group O
   (c)   All groups

5. The red blood cells of the fetus may haemolyse if the mother has formed antibodies to the Rh^+ cells.

7. THE LYMPHATIC SYSTEM

7.1. Lymphatic capillaries and vessels

The blood capillaries are impermeable to large molecules such as proteins. Any protein which leaks into the tissue fluid from cells or plasma can not therefore enter the blood stream.

These molecules are absorbed, along with some tissue fluid, by the lymphatic system.

Lymphatic capillaries are blind-ended and thin-walled. They are found in all the tissues of the body except those of the central nervous system.

The capillaries unite to form lymphatic vessels which are thin walled, and have large numbers of valves.

They all drain into lymph nodes which filter the lymph and pass it on to further vessels.

The lymph vessels of the upper right part of the body finally drain into the right lymphatic duct which empties the lymph into the right subclavian vein.

The vessels from the rest of the body join to form the thoracic duct which empties into the left subclavian vein.

The lymph in the thoracic duct contains a high proportion of fats absorbed from the villi of the small intestine.

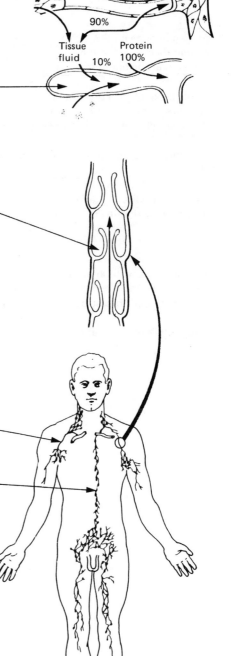

Capillary

90%

Tissue fluid

10%

Protein 100%

7.2. Lymph nodes

Lymph nodes (or glands) are masses of lymphoid tissue which lie along the course of lymphatic vessels and which filter the lymph. They vary in size from 1—10 mm and are most numerous in the axillae, groin, neck and posterior wall of the abdominal cavity.

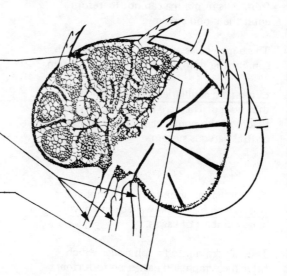

Incoming lymph empties into a subscapular sinus and then filters to the outgoing efferent lymphatics at the hilum.

The substance of the node is built on a framework of reticulin fibres with endothelial cells forming perforated tubes.

Within this there are masses of lymphocytes, at the centre of which lie germinal centres. These contain the plasma cells which produce antibodies. Lymphocytes are mainly produced in the lymph nodes.

Lymphoid tissue without a capsule forms the tonsils and adenoids, and is found in patches throughout the small and large bowel, including the appendix. All of these sites, since they filter and attempt to destroy bacteria, are liable to infection.

The total bulk of lymphatic tissue in the body is approximately equal to that of the liver.

7.3. The spleen

The spleen is a soft organ which lies on the left side of the abdomen, under cover of the ribs, immediately beneath the diaphragm.

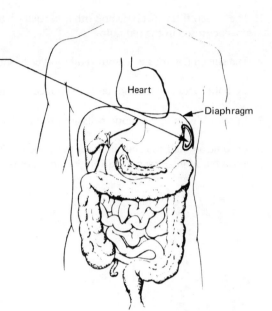

Heart

Diaphragm

It weighs about 200 g and is about 125 mm long. It cannot usually be felt through the abdominal wall, but may become greatly enlarged in certain diseases.

It consists of a mass of red pulp with millions of pin-head nodules of white pulp scattered through it giving it a granular appearance.

The spleen has a rich blood supply via the splenic artery. The blood drains into the portal vein.

Blood entering the spleen via the artery is filtered. The tiny branches of the artery when they leave the fibrous framework first pass through a nodule of white pulp.

The blood then flows in arterioles through masses of reticulo-endothelial cells which are highly phagocytic (that is they ingest micro-organisms, other cells and foreign particles).

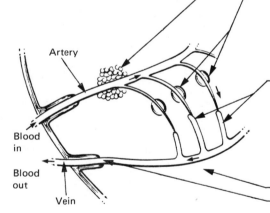

Artery

Blood in

Blood out

Vein

The blood finally drains into venous sinuses.

Blood leaks from these venous sinuses and circulates freely through the red pulp before finally draining into the main vein.

The cells of the body which, like those of the spleen and lymph node, are able to ingest foreign particles are sometimes said to belong to the *reticulo-endothelial system*.

7.4. The functions of the spleen

1. The spleen filters bacteria and other particles from the blood stream. These particles are phagocytosed in the red pulp.

2. The spleen destroys worn-out red blood cells and platelets.

3. Lymphocytes and monocytes are produced in the white pulp of the spleen.

4. In fetal life only, red blood cells are produced in the spleen.

5. In some animals (but not in man) the spleen acts as a reservoir for blood and can release extra blood into the circulation during times of stress.

TEST SEVEN

1. **Are the following statements true or false?**

		True	False
(a)	Lymphatic capillaries are found in all the tissues of the body.	( )	( )
(b)	Lymphatic capillaries are found in nearly all the tissues of the body.	( )	( )
(c)	Lymphatic capillaries are found in nervous tissue.	( )	( )
(d)	Lymphatic capillaries are found in muscular tissue.	( )	( )

2. **Indicate which of the following are functions of lymph nodes and which are functions of the spleen by placing ticks in the appropriate boxes.**

		Lymph nodes	Spleen
(a)	Filtration of bacteria and other particles from the lymph.	( )	( )
(b)	Filtration of bacteria and other particles from the blood stream.	( )	( )
(c)	Production of lymphocytes.	( )	( )
(d)	Destruction of red blood cells and platelets.	( )	( )
(e)	Acts as a blood reservoir.	( )	( )

ANSWERS TO TEST SEVEN

1.

		True	False
(a)	Lymphatic capillaries are found in all the tissues of the body.	( )	(✓)
(b)	Lymphatic capillaries are found in nearly all the tissues of the body.	(✓)	( )
(c)	Lymphatic capillaries are found in nervous tissue.	( )	(✓)
(d)	Lymphatic capillaries are found in muscular tissue.	(✓)	( )

2.

		Lymph nodes	Spleen
(a)	Filtration of bacteria and other particles from the lymph.	(✓)	( )
(b)	Filtration of bacteria and other particles from the blood stream.	( )	(✓)
(c)	Production of lymphocytes.	(✓)	(✓)
(d)	Destruction of red blood cells and platelets.	( )	(✓)
(e)	Acts as a blood reservoir.	( )	( )

POST TEST

1. **Indicate which of the names in the list below refer to the arteries labelled on the diagram alongside, by placing the appropriate letters in the boxes.**

   1. Common iliac artery ( )
   2. Superior mesenteric artery ( )
   3. Common carotid artery ( )
   4. Abdominal aorta ( )
   5. Femoral artery ( )
   6. Brachial artery ( )
   7. Pulmonary artery ( )

   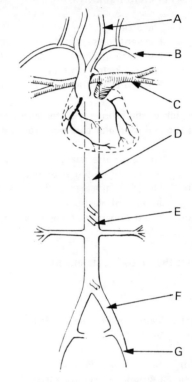

2. **Indicate which of the names in the list below refer to the parts of the heart labelled on the diagram alongside by placing the appropriate letters in the boxes.**

   1. Left ventricle ( )
   2. Right atrium ( )
   3. Aorta ( )
   4. Pulmonary artery ( )
   5. Pulmonary vein ( )
   6. Inferior vena cava ( )
   7. Semilunar valve ( )
   8. Tricuspid valve ( )

   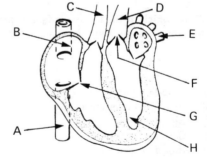

3. **The diagram alongside is a representation of an electrocardiogram.**

   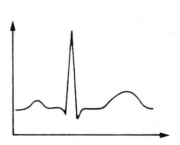

   Label:
   (a) the P wave
   (b) the QRS wave
   (c) the T wave

   Indicate the points at which the electrocardiogram shows the activity caused by:

   (d) the contraction of the ventricles
   (e) the contraction of the atria.

4. **Place the following events in the correct order to describe the cardiac cycle in outline. Start at event number (i).**

(i)	The atria fill	(vii)	The atria contract
(ii)	The ventricles empty	(viii)	The ventricles relax
(iii)	The semilunar valves open	(ix)	The atrio-ventricular valves close
(iv)	The ventricles contract	(x)	The semilunar valves close
(v)	The ventricles fill	(xi)	The ventricles are topped up

5. **Which of the following statements correctly represents the relative proportions of elastic tissue and muscular tissue in large and small arteries?**

(a)   The larger the artery, the larger the amount of elastic tissue and the smaller the amount of muscular tissue.

(b)   The larger the artery, the larger the amount of muscular tissue and the smaller the amount of elastic tissue.

(c)   The proportion of elastic and muscular tissue is the same in both large and small arteries.

6. **Are the following statements true or false?**

		True	False
(a)	Veins have thicker walls than arteries.	( )	( )
(b)	Veins have thinner walls than arteries.	( )	( )
(c)	Veins have valves.	( )	( )
(d)	Arteries have valves.	( )	( )
(e)	Lymph vessels have valves.	( )	( )
(f)	Pressure in arteries and veins is much the same.	( )	( )
(g)	Arterial pressure is greater than venous pressure.	( )	( )

7. **Indicate which of the following factors are usually variable and which are usually constant in a normal healthy adult, by placing ticks in the appropriate boxes.**

		Variable	Constant
(a)	Blood viscosity	( )	( )
(b)	Peripheral resistance	( )	( )
(c)	Blood volume	( )	( )
(d)	Cardiac output	( )	( )
(e)	Blood pressure	( )	( )

8. **Indicate which of the following factors cause arteriolar dilation and which cause constriction by placing ticks in the appropriate boxes.**

		Causes arteriolar dilation	Causes arteriolar constriction
(a)	Lack of oxygen	( )	( )
(b)	Excess carbon dioxide	( )	( )
(c)	Histamine	( )	( )
(d)	Noradrenaline	( )	( )

9. **What volume of the blood is occupied by plasma?**

   (a) 25%   (b) 35%   (c) 45%   (d) 55%   (e) 65%   (f) 75%

10. **Place the following events in the order in which they occur in haemostasis.**

   (i)   Fibrinogen ⟶ fibrin
   (ii) Prothrombin ⟶ thrombin
   (iii) Platelets adhere to injured tissue
   (iv) Release of serotinin
   (v)   Formation of a solid clot
   (vi) Release of exudate

11. **Which of the blood groups listed on the left can be safely used in transfusions to persons of the blood types listed on the right?**

   (i)   Blood of group O   (a) Group O recipient
   (ii) Blood of group A   (b) Group A recipient
   (iii) Blood of group B   (c) Group B recipient
   (iv) Blood of group AB   (d) Group AB recipient

12. **Indicate which of the names in the list below refer to the parts of the lymph node labelled on the diagram alongside by placing the appropriate letters in the boxes.**

   1. Subscapular sinus   ( )
   2. Germinal centre   ( )
   3. Efferent lymphatics   ( )

ANSWERS TO POST TEST

1.
1. Common iliac artery (F)
2. Superior mesenteric artery (E)
3. Common carotid artery (A)
4. Abdominal aorta (D)
5. Femoral artery (G)
6. Brachial artery (B)
7. Pulmonary artery (C)

2.
1. Left ventricle (H)
2. Right atrium (B)
3. Aorta (D)
4. Pulmonary artery (C)
5. Pulmonary vein (E)
6. Inferior cena cava (A)
7. Semilunar valve (F)
8. Tricuspid valve (G)

3.

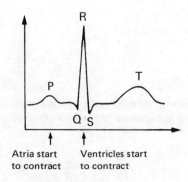

Atria start to contract

Ventricles start to contract

4.
1. (i) The atria fill
2. (vi) The atrio-ventricular valves open
3. (v) The ventricles fill
4. (vii) The atria contract
5. (xi) The ventricles are topped up
6. (iv) The ventricles contract
7. (ix) The atrio-ventricular valves close
8. (iii) The semilunar valves open
9. (ii) The ventricles empty
10. (x) The semilunar valves close
11. (viii) The ventricles relax

5. (a) The larger the artery, the larger the amount of elastic tissue and the smaller the amount of muscular tissue.

6.

		True	False
(a)	Veins have thicker walls than arteries.	( )	(✓)
(b)	Veins have thinner walls than arteries.	(✓)	( )
(c)	Veins have valves.	(✓)	( )
(d)	Arteries have valves.	( )	(✓)
(e)	Lymph vessels have valves.	(✓)	( )
(f)	Pressure in arteries and veins is much the same.	( )	(✓)
(g)	Arterial pressure is greater than venous pressure.	(✓)	( )

7.

		Variable	Constant
(a)	Blood viscosity	( )	(✓)
(b)	Peripheral resistance	(✓)	( )
(c)	Blood volume	( )	(✓)
(d)	Cardiac output	(✓)	( )
(e)	Blood pressure	(✓)	( )

8.

		Causes arteriolar dilation	Causes arteriolar constriction
(a)	Lack of oxygen	(✓)	( )
(b)	Excess carbon dioxide	(✓)	( )
(c)	Histamine	( )	(✓)
(d)	Noradrenaline	( )	(✓)

9. (d) 55%

10.
1. (iii) Platelets adhere to injured tissue
2. (iv)   Release of serotinin
3. (ii)   Prothrombin → thrombin
4. (i)   Fibrinogen → fibrin
5. (v)   Formation of a solid clot
6. (vi)   Release of exudate

11.
(a) Group O recipient can receive (i) blood of group O only.
(b) Group A recipient can receive (i) blood of group O and (ii) blood of group A.
(c) Group B recipient can receive (i) blood of group O and (iii) blood of group B.
(d) Group AB receipient can receive (i) (ii) (iii) (iv) blood of all groups.

12.
1. Subscapular sinus   (A)
2. Germinal centre   (B)
3. Efferent lymphatics   (C)